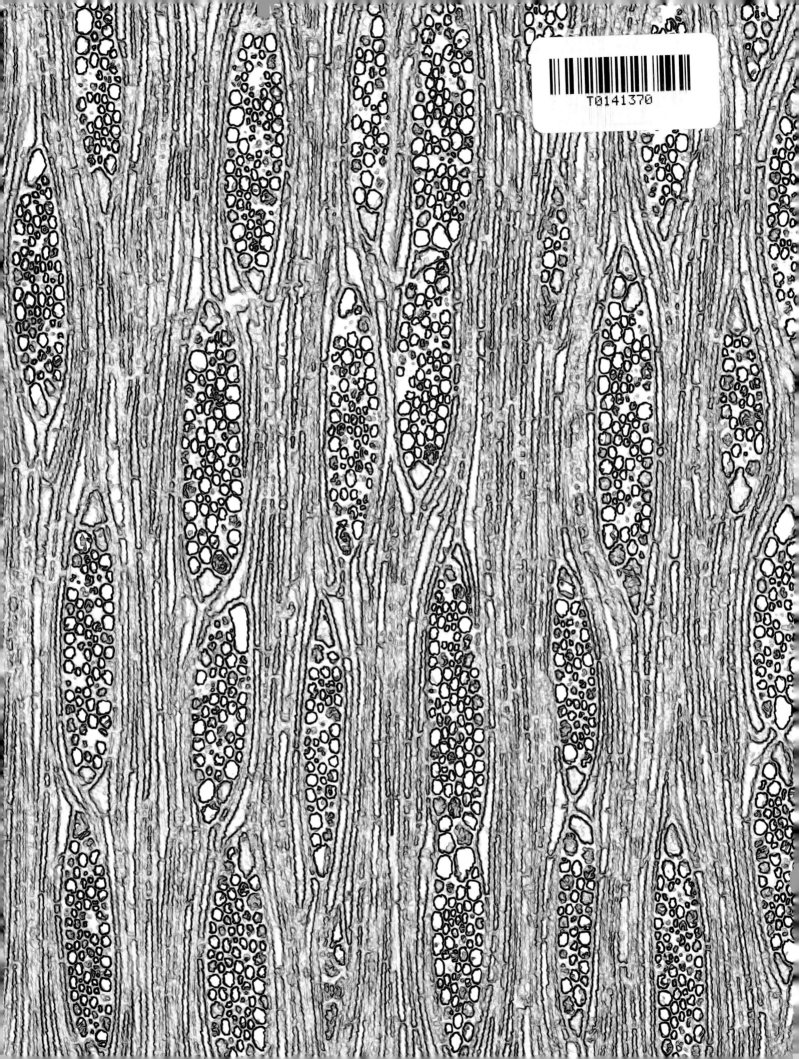

# Mahogany

This book is dedicated to the late Donald and Margaret Gasson,
whose support and guidance will be greatly missed.

# Mahogany

Lydia White and Peter Gasson

Kew Publishing
Royal Botanic Gardens, Kew

PLANTS PEOPLE
POSSIBILITIES

First published in 2008 by
Royal Botanic Gardens, Kew
Richmond, Surrey, TW9 3AB, UK
www.kew.org

ISBN 978 1 84246 171 6

British Library Cataloguing in Publication Data
A catalogue record for this book is available from the British Library

Production Editor: Michelle Payne
Typesetting and page layout: Christine Beard
Design by Media Resources, Royal Botanic Gardens, Kew

Front cover photograph: *Swietenia mahagoni*, longitudinal surface (Andrew McRobb)
Frontispiece: *Swietenia mahagoni* tree in the Florida Everglades (Peter Gasson)
End papers: *Swietenia macrophylla*, tangential longitudinal surface (Lydia White)

All other photographs by Andrew McRobb/Lydia White © The Board of Trustees of the Royal Botanic Gardens, Kew 2008

Printed in Italy by Printer Trento

For information or to purchase all Kew titles please visit
www.kewbooks.com or email publishing@kew.org

All proceeds go to support Kew's work in saving the world's plants for life

# CONTENTS

# ACKNOWLEDGEMENTS

We would like to thank Terry Pennington for his help and advice, Stephen Harris (Oxford University) for the loan of mahogany woods from the Xylarium at Oxford University, and Regis Miller (Forest Products Laboratory, Madison, Wisconsin) for wood samples. We have many people to thank here at Kew Gardens: Andrew McRobb for the macroscopic images of the mahogany woods; Gina Fullerlove, John Harris, Michelle Payne, Lloyd Kirton and Christine Beard from Kew Publishing for their work and support throughout this project; Susana Baena for the maps; Trevor Beckham for sanding the woods; Noel McGough and all in Kew's Conventions and Policies Section for their consistent support in CITES wood projects; and Paula Rudall and all in the Micromorphology section of the Jodrell laboratory for supporting this project.

# INTRODUCTION

## About the book

Mahogany is one of the finest cabinet-making timbers in the world. Over-exploitation is now driving mahogany (*Swietenia* species) towards extinction, and there are international laws in place to control its trade. With legislation comes the need to accurately identify and separate *Swietenia*, the true mahogany, from other similar-looking timbers that could be confused with it. *Swietenia* belongs to the family Meliaceae (often known as the mahogany family), which comprises 52 genera and 621 species (Stevens, 2001 onwards), many of which are timbers of commercial value. A number of other genera in the Meliaceae closely resemble *Swietenia*, as do various non-Meliaceae timbers.

This book is a guide to the macroscopic appearance and the microscopic structure of mahogany and similar woods. It aims to be scientifically accurate and aesthetically pleasing, and to appeal to a wide audience, from front-line customs officers, furniture lovers, to those with a more scientific interest in timber.

This book is divided into several sections. The introductory pages provide a brief history of mahogany, the origins of its name, and its modern status. The main body of the book gives detailed descriptions of 18 genera of mahogany timbers in the family Meliaceae, each described in four pages with information on the natural distribution, uses, common names, wood anatomy (including diagnostic characteristics) and full-colour macroscopic and microscopic images of the wood. These accounts are designed to enable the identification of *Swietenia* and other similar mahogany woods.

The Meliaceae woods included in this book were chosen because they are generally traded as 'mahogany', as well as under many other common names. These timbers share similarities in colour and in wood anatomy with *Swietenia*. The common names given in each entry are not exhaustive. Those used in international trade are given and so are some local names.

The mahogany woods have been arranged in alphabetical order. Of the taxa included, *Swietenia*, *Khaya* and *Entandrophragma* are considered to be the most valuable timbers of the family (Miller, 1990). Where particular species have been included in genus accounts, these have been chosen as representatives of the genus, often based upon their commercial importance. Some of the taxa included in this section are insignificant in international trade, but their use in the future may increase as stocks of true mahogany in the wild decline, and the need to use other mahogany-like woods becomes apparent.

A large number of non-Meliaceae woods are also known as 'mahogany' and they are considered more briefly in a table, following the main section of the book. This table lists the Latin and common names of the timbers, geographical distribution, and the anatomical characteristics of the wood that distinguish it from *Swietenia*, along with full-colour anatomical images. This is designed as a quick-reference section, as the wood anatomy of these species is so different from *Swietenia* and full accounts would have greatly lengthened the book.

The appendices include details of the microscope slides studied and timber trade data between 1996–2005 for *Swietenia*. Finally, there is a list of the literature used in the production of this book, which will provide further and detailed information to the reader.

The wood anatomical descriptions follow the definitions of characters detailed in the IAWA list of microscopic features for hardwood identification (Wheeler, Baas & Gasson, 1989). There are too many features to define here, but a few conventions and observations have been used:

TS = Transverse Section (i.e. Cross Section)

TLS = Tangential Longitudinal Section

RLS = Radial Longitudinal Section

Intervessel pits described as minute are less than, or equal to 4 μm in diameter; small intervessel pits are 4–7 μm in diameter, medium intervessel pits are 7–10 μm in diameter, and the diameter of large intervessel pits is 10 μm or greater.

The information presented in this book has come from diverse and extensive sources, from original observations to general references that include Record and Hess (1943), Pennington and Styles (1975, 1981), Mabberley (1997), and Walker (2005) and more specific references including Groom (1926), Kribs (1930), Panshin (1933), Rendle (1938), Kribs (1959), Ghosh *et al.* (1963), British Standards Institution (1991), Wong (2002), and the InsideWood website (2004 onwards). For full details of all the references used in this book, see the Further Reading list.

# A brief history of mahogany

*Swietenia* mahogany is a popular and valuable wood, and has been harvested for centuries. The genus is composed of three species, *Swietenia macrophylla*, *Swietenia mahagoni*, and *Swietenia humilis*. *Swietenia* was named after the Dutch botanist and physician Gerard von Swieten (1700–1772) (Miller, 1990) by another Dutch botanist, Nikolaus Joseph von Jacquin, who described *S. mahagoni* in 1760 (Lamb, 1966). This name was based on *Cedrela mahagoni*, originally proposed by Linnaeus in 1759. The use of the species name *mahagoni* gave status to the common name 'mahogany' which was already in widespread use (Lamb, 1966).

The origins of the name 'mahogany' are unclear, but may relate to a generic name for African mahogany, Oganwo, which had been used by Nigerian slaves in the Caribbean (Lamb, 1966; Pennington & Styles, 1981). *Swietenia humilis* was described by J. G. Zuccarini between 1836 and 1837, and *S. macrophylla* was described in 1886 by Sir George King (Record & Hess, 1943). All other 'species' of *Swietenia* are now considered to belong to the three species already mentioned.

The use of mahogany dates back to the 1500s, when *S. mahagoni* was originally used by the Spanish in the Caribbean for the construction of ships and buildings, as the wood had great resistance to dry rot, termites and warping (Miller, 1990). The introduction of mahogany in England has been associated with Sir Walter Raleigh, who is said to have presented a mahogany table to Queen Elizabeth I (Record & Hess, 1943). In 1680, mahogany was used for flooring and choir stalls in Nottingham Castle (Record & Hess, 1943). By the early 1700s, the name 'mahogany' was in common use (Lamb, 1966), and the wood became popular (and still is) for furniture and cabinet making, due to its superior working and finishing properties, and attractive colour and grain (Rendle, 1969). Trade in mahogany flourished in the latter half of the 1800s following the removal of high import duties in the mid-1800s (Melville, 1936).

Early harvesting of mahogany began with *S. mahagoni* (Caribbean mahogany) in the seventeenth and eighteenth centuries (Record, 1924), from the Caribbean and the southern tip of Florida (Record & Hess, 1943). At the time, *S. mahagoni* was prized as one of the most valuable and attractive commercial timbers on the market (Lincoln, 1986). The wood of *S. mahagoni* is considered superior to *S. macrophylla* due to its finer grain and more attractive finished colour (Miller, 1990). *Swietenia humilis* (Honduras mahogany) occurs in dry areas along the Pacific coast of Mexico and Central America (Record & Hess, 1943) and used to be harvested along with *S. mahagoni*, but by no means to the same extent. The geographical distribution of these two species meant that they were more accessible than timbers in the heart of South America (such as *S. macrophylla*), and therefore became over-logged. Both species are now considered by CITES (see next section) as commercially extinct.

As the popularity and demand for mahogany wood grew, the supplies began to shrink, and alternatives were sought. This led to the establishment of *S. macrophylla* plantations in India and South-East Asia, and the increasing use of mahogany-like woods, such as the West African timbers *Khaya* and *Entandrophragma* (Melville, 1936).

*Swietenia macrophylla* (Bigleaf mahogany) is not yet commercially extinct, perhaps due to its wide natural distribution which extends from the southern end of Mexico, through Central America and into South America (Colombia, Venezuela, Peru, Bolivia, and Brazil), in areas of heavy rainfall (Record & Hess, 1943), more specifically in strongly seasonal climates and much less so in everwet forest (Pennington, 2002). Populations of *S. macrophylla* still remain, mainly in South America, but over-harvesting and illegal logging endanger its survival. In addition, the land that *S. macrophylla* grows on is under threat as it is highly valued for agriculture and livestock grazing (Blundell, 2004).

Logging can be detrimental to *Swietenia*. It can take 52 years for a tree to reach commercial size at maximum growth rate, and 148 years at medium growth rate (Gullison *et al.*, 1996), but under the right conditions it can be fast growing (Pennington, pers. comm.). *Swietenia macrophylla* has the ability to withstand hurricanes, flooding, and fires, and is able to disperse seeds and grow without much competition from neighbouring trees (Snook, 1996). However, as a consequence of heavy logging, adult *S. macrophylla* trees are removed with few seeds remaining, resulting in a population decrease (Snook, 1996). Although *Swietenia* plantations have been established, their success can be limited by attack from *Hypsipyla* shoot borers (moth larvae) (Miller, 1990).

For more detailed historical accounts of mahogany, see Record and Hess (1943), Lamb (1966), and Pennington and Styles (1981).

## Mahogany and CITES

All three of the *Swietenia* species are listed in Appendix II (EU Annex B) of the Convention on International Trade of Endangered Species of wild fauna and flora (CITES) (see www.cites.org). There are three CITES Appendices (Appendix I has the tightest restrictions on trade), and species are listed depending upon the threat of extinction. The Appendix II listing of *Swietenia* means that these species may become threatened with extinction, and therefore trade is only allowed subject to a permit.

*Swietenia humilis* has been listed on Appendix II since 1975 and *S. mahagoni* since 1992 (www.cites.org), as they have long been considered commercially extinct. After years of campaigning from various organisations and CITES parties (countries), *S. macrophylla* was moved from Appendix III (listed since 1995) to Appendix II (in November 2003). The CITES listing of *S. macrophylla* covers logs, sawn wood, veneer sheets, and plywood from all neotropical nations (Central and South America, and the Caribbean), but not from plantations elsewhere such as in South-East Asia (Soerianegara & Lemmens, 1993), or antiques. In addition to the CITES listing, *Swietenia* is also on the IUCN Red List of Threatened Species, which classifies species at a high risk of global extinction (*see* www.iucnredlist.org). Many of the Meliaceae species included in this book are on the IUCN Red List (*Aglaia argentea, Aphanamixis polystachya, Cedrela fissilis, Cedrela odorata, Entandrophragma, Guarea, Khaya, Lovoa, Swietenia, Toona ciliata* and *Turraeanthus africanus*).

# Timber identification

Identification and separation of *Swietenia* from other types of mahogany is difficult because of the close similarities between the colour and grain of the woods. The attractive red/brown colour of *Swietenia* is characteristic of many other woods, both within and outside the family Meliaceae. Completely unrelated woods can often resemble *Swietenia*, such as many Dipterocarpaceae from South-East Asia. Fortunately, the wood anatomy of non-Meliaceae species differs sufficiently from *Swietenia* and other Meliaceae mahoganies, so telling them apart microscopically is not difficult. However, the Meliaceae mahoganies are often almost identical macroscopically, and very similar microscopically.

To add to this confusion, variations in timber colour and uniformity of wood characteristics occur within different samples of the same species, which is often a consequence of the site and conditions of growth (Record & Hess, 1943). Variation in wood samples has led to discoveries of 'new' *Swietenia* species, such as *S. krukovii* (Gleason & Panshin, 1936), *S. cirrhata, S. candollei* and *S. tessmannii*, none of which are now recognised as species (Record & Hess, 1943).

The wood of the three species of *Swietenia* is very difficult to separate anatomically (Panshin, 1933; Ghosh *et al.*, 1963). The species 'are poorly defined biologically probably because they hybridise freely' when their largely allopatric distributions come into contact (Pennington & Styles, 1981), creating intermediate morphological traits (Miller, 1990). Timber from the genus *Khaya*, known as African mahogany, has long been considered the most similar to *Swietenia*. Not only does the wood of *Khaya* resemble *Swietenia* macroscopically, with its red/brown colouring and similar grain, it also resembles the true mahogany microscopically (Kribs, 1930; Panshin, 1933). This similarity is not surprising as the two genera are very closely related (Chalmers *et al.*, 1994; Muellner *et al.*, 2003).

The aim of this book is to act as an identification guide to the Meliaceae mahoganies, with the primary goal of enabling the reader to successfully recognise the endangered *Swietenia* species, especially in view of its legal protection.

# MELIACEAE MAHOGANIES

## Guide to the Meliaceae mahogany accounts

Each account comprises four pages.

**Genus/species examined:** all *Swietenia* species and those of monotypic genera are considered. For larger genera only the species most likely to be encountered are referred to. The authors who first described the taxon are given.

**Introduction:** a brief account putting the species examined in the context of the genus.

**Natural distribution:** the countries or general region in which a species naturally occurs are given, and where appropriate plantations outside the natural distribution are mentioned.

**Map:** the outline of the map is the same for all taxa, and shows only the countries in which the species occur naturally. The distributions are based upon the 2006 IUCN Red List of Threatened Species (www.iucnredlist.org), Pennington and Styles (1975, 1981), Phongphaew (2003), Lemmens *et al.* (1995), Soerianegara and Lemmens (1993), and Sosef *et al.* (1998).

**Wood:** the colour of the wood and macroscopic features are given here.

**Uses:** the main uses of the timber are listed.

**Macroscopic photographs:** two photographs are given, showing the transverse surface and a longitudinal surface. The latter is not consistent between species and depends on the plane of the surface on the sample. Where necessary, the images have been magnified to accommodate wood samples of differing sizes. This is indicated in the figure caption (for example as × 2). However, in all cases a piece of the longitudinal surface is also reproduced life size, which can be very helpful whan matching a piece of wood with the photograph.

**Common names:** the most frequently used commercial and vernacular names are given. Some species have so many names that not all are listed.

**Anatomy:** this describes the wood anatomical characters in the following order: growth rings, porosity, vessels, fibres, axial parenchyma, rays and cellular inclusions (crystals and silica bodies). References are given to descriptions and photographs in the literature.

**Diagnostic characteristics:** summarises the few wood anatomical characters that appear to be diagnostic for particular genera or species.

**Figures:** these show TS, TLS and RLS photographs of the salient features of each genus. The colour of the sections is a result of chemical staining.

# *Aglaia argentea* Blume

## Introduction

There are over 100 species in the genus *Aglaia*, distributed across India, South-East Asia and Australia. The morphology of this genus is variable due to the large number of species, including the former genus *Amoora*, which was sunk into *Aglaia*. *Aglaia argentea* is a commercially important timber often used as a mahogany substitute.

## Natural distribution

*Aglaia argentea* is found in Myanmar (Burma) and is extensively distributed across South-East Asia, including Thailand, Malaysia, Indonesia and Australia.

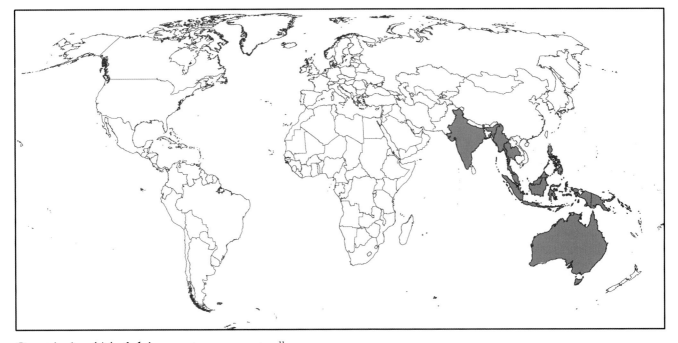

Countries in which *Aglaia argentea* occurs naturally

## Wood

As with most mahogany woods, the colour of the timber varies between light and dark red/brown.

## Uses

The timber is used locally, mostly in construction, furniture, and flooring. Due to its attractive appearance the timber is also used for high-quality cabinet work, as are many of the mahogany woods. Heavier timbers of the genus are used in the construction of houses and bridges.

## Common names

General names for *Aglaia* include Pasak, Amoora, Bekak, Langsat, Segera and Tasua. The common names vary between regions and countries, as they are often based on local languages.

***Aglaia argentea*** transverse surface, × 5

***Aglaia argentea*** longitudinal surface, × 1 (inset) and × 1.5

## Anatomy

Figures A–E

Growth ring boundaries indistinct or absent. Wood diffuse-porous. Vessels solitary and in radial multiples (Fig. A) and clusters of up to five. Simple perforation plates. Intervessel pits minute and alternate (Fig. C). Vessel-ray pits with distinct borders and similar in size and shape to intervessel pits. Fibre walls thin to thick. Septate fibres present (Fig. E). Paratracheal parenchyma occasionally scanty, winged aliform and confluent (Fig. A). Rays not storied, uniseriate (to biseriate) and 10–30 cells high (Fig. B). Rays heterocellular with one row of square and/or upright marginal cells (Fig. D). Abundant prismatic crystals present in chambered cells in chains, in the axial parenchyma (Fig. E).

References to the wood anatomy of *Aglaia* include Kribs (1930), Metcalfe and Chalk (1950), Desch (1954), Ghosh *et al.* (1963), Pennington and Styles (1975), Wong (1976), Datta and Samanta (1983), Lemmens *et al.* (1995) and Negi *et al.* (2003).

## Diagnostic characteristics

The winged aliform and confluent parenchyma distinguish *A. argentea* from *Swietenia* (pp. 68–69). In addition, the tall uniseriate rays in *A. argentea* are not commonly seen in other Meliaceae woods.

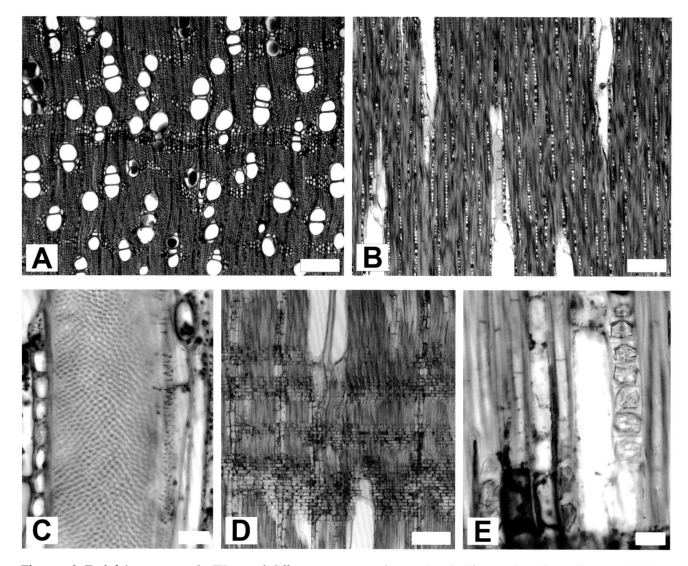

**Figures A–E** *Aglaia argentea*. **A.** TS, wood diffuse porous, parenchyma winged-aliform and confluent, forming indistinct bands, also scanty paratracheal. **B.** TLS, rays tall and uniseriate, not storied. **C.** TLS, intervessel pits alternate and minute. **D.** RLS, rays heterocellular. **E.** RLS, chain of prismatic crystals in axial parenchyma. Septate fibres also present. Figs. A, B and D, scale bar = 200 μm, Figs. C and E, scale bar = 20 μm.

# *Aphanamixis polystachya* (Wall.) R. Parker

## Introduction

There are three or four species of *Aphanamixis*, but *A. polystachya* is the only one used as a commercial timber. Historically, this South-East Asian timber has had various Latin names attributed to it, including *Amoora aphanamixis*, *Aphanamixis rohituka* and *Aphanamixis grandiflora*.

## Natural distribution

*Aphanamixis polystachya* is distributed across South and South-East Asia, including India, Sri Lanka, Bhutan, Myanmar (Burma), Thailand, Malaysia and Vietnam.

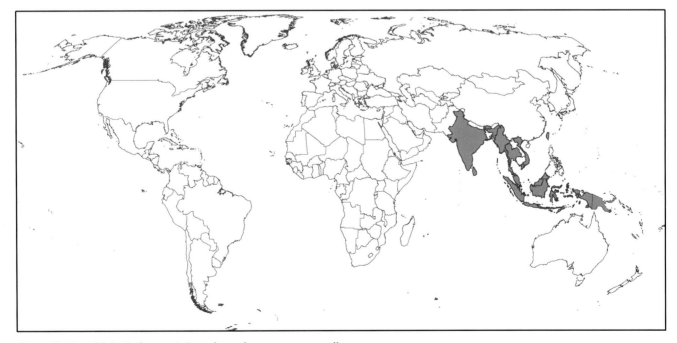

Countries in which *Aphanamixis polystachya* occurs naturally

## Wood

The timber has the attractive appearance of true mahogany, as it is a pale to dark red shade of brown.

## Uses

The wood is used for house construction, furniture and boat production, and as a veneer. The tree is planted as an ornamental or shade tree, and is often utilised for seed oil and soaps. Bark extracts are used for medicines.

## Common names

Depending upon locality, language and trade, *A. polystachya* is commonly known as Tasua, Kayu gula, Chikek and Salakin.

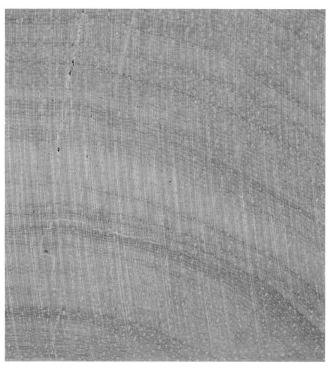

*Aphanamixis* species transverse surface, × 3.5

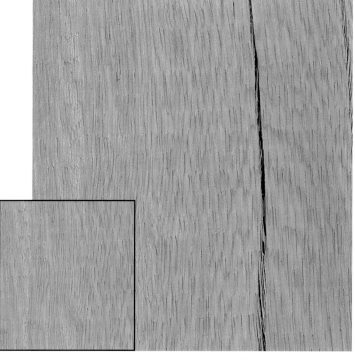

*Aphanamixis* species longitudinal surface × 1 (inset) and × 1.5

## Anatomy

Figures A–D

Growth ring boundaries indistinct or absent. Wood diffuse-porous. Vessels solitary and in radial multiples and clusters of up to five (Fig. A). Simple perforation plates. Intervessel pits small and alternate (Fig. C). Vessel-ray pits with distinct borders and similar in size and shape to intervessel pits. Fibre walls thin to thick. Septate fibres present. Paratracheal parenchyma scanty to winged aliform and confluent, forming indistinct, wavy bands 3–5 cells wide (Fig. A). Rays not storied, uniseriate and biseriate, and up to 20 cells high (Fig. B). Rays heterocellular with 1–2 rows of square and/or upright marginal cells (Fig. D). Prismatic crystals present in chambered axial parenchyma cells. Silica bodies present in ray and axial parenchyma cells.

References to the wood anatomy of *Aphanamixis* include Kribs (1930), Metcalfe and Chalk (1950), Desch (1954), Pennington and Styles (1975), Sosef *et al.* (1998), and the InsideWood website (2004 onwards).

## Diagnostic characteristics

The parenchyma of *A. polystachya* is arranged in irregular, wavy bands (similar to *Guarea* (Fig. A, p. 49)), which is clearly different from the continuous bands of parenchyma found in *Swietenia*. The rays are also different as they are uniseriate and biseriate, and not storied. The chains of crystals that are common in other Meliaceae mahoganies, but not in *Swietenia*, are found in *A. polystachya*, as are silica bodies (also found in *Entandrophragma candollei* (p. 44), *Guarea* (p. 48) and *Chisocheton divergens* (pp. 32–33)).

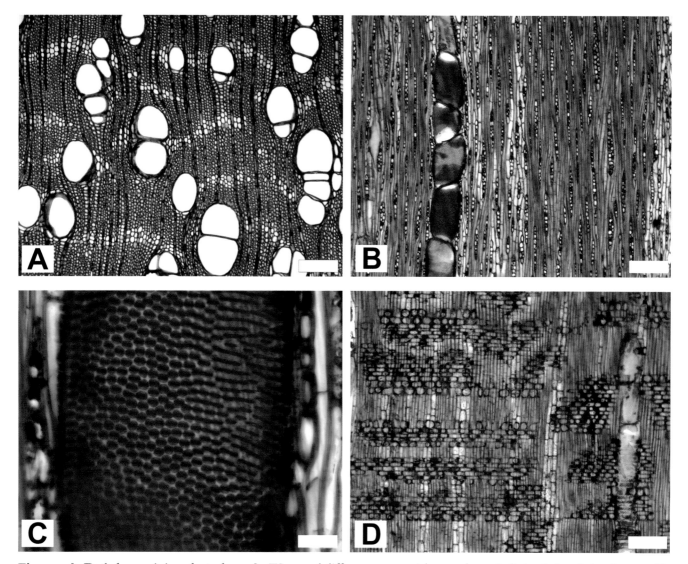

**Figures A–D** *Aphanamixis polystachya.* **A.** TS, wood diffuse porous, axial parenchyma indistinctly banded and wavy. **B.** TLS, rays uniseriate and biseriate, fibres septate. **C.** TLS, intervessel pitting alternate and small. **D.** RLS, rays heterocellular. Figs. A, B, D, scale bar = 200 μm, Fig. C = 20 μm.

# *Azadirachta* A.Juss.

## Introduction

There are two species of *Azadirachta*, both of which are internationally traded and resemble mahogany wood. *Azadirachta excelsa* and *A. indica* originate from Asia and South-East Asia, and *A. indica* in particular is very widely used and has many practical applications. In the past, because of the similarities between *Azadirachta* and *Melia*, names of the two genera have been confused. For example, *A. indica* has been described under the names *Melia indica* and *Melia azadirachta*, which should not be confused with *Melia azedarach* (pp. 58–61).

## Natural distribution

The distribution of *Azadirachta* is similar to that of *Melia*, across Asia and South-East Asia. Plantations of *A. indica* exist in South-East Asia, Africa, North and South America, Cuba and Central America.

***Azadirachta excelsa*** occurs in Peninsular Malaysia, Sumatra, Borneo, Sulawesi, the Aru Islands, Papua New Guinea and the Philippines.

***Azadirachta indica*** occurs in India and Myanmar (Burma).

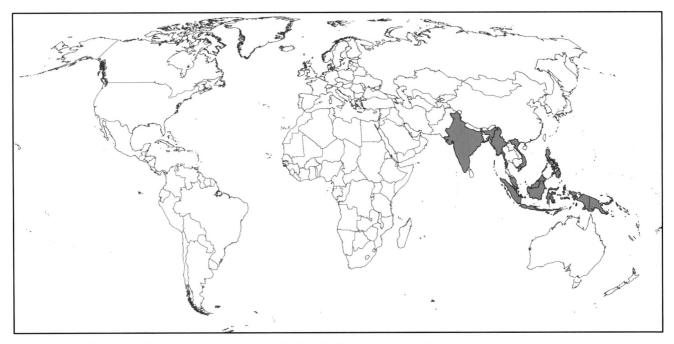

Countries in which ***Azadirachta excelsa*** and ***Azadirachta indica*** occur naturally

# *A. excelsa* (Jack) Jacobs and *A. indica* A.Juss.

## Wood

The wood is pale to dark brown/red, similar to *Melia* and many other Meliaceae mahoganies. Growth rings and occasional traumatic resin canals are sometimes visible to the naked eye on the transverse surface of the wood.

## Uses

Both species of *Azadirachta* are used locally and internationally in light construction, furniture production, panelling, and even as fuel wood (*A. indica*). *Azadirachta indica* is used more as a shade and ornamental tree rather than primarily as timber, and many products are made from the flowers and bark (such as insect repellents).

## Common names

The wide distribution and plantations of *Azadirachta* around the world has led to the existence of many common names. *Azadirachta excelsa* is known as Sentang, Kayu bawang, Marango, and Bird's Eye Kalantas. *Azadirachta indica* is famously known as Neem, but may also be called Nim, Margosa, Khwinim, or Imba.

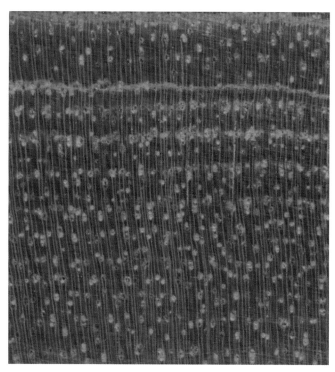

*Azadirachta indica* transverse surface, × 5.5

*Azadirachta indica* longitudinal surface, × 1 (inset) and × 2

## Anatomy

Figures A–G

Growth rings present. Diffuse-porous wood. Vessels solitary and in radial multiples and clusters up to six. Simple perforation plates. Alternate intervessel pitting minute to small (Fig. E). Vessel-ray pits with distinct borders and similar in size and shape to intervessel pits. Fibre walls thin to thick. Striations in vessel walls resembling helical thickenings sometimes visible. All fibres non-septate. Banded axial parenchyma, 3–6 cells wide (Figs. A and B). Scanty (to vasicentric) paratracheal axial parenchyma (Figs. A and B). Rays not storied, 2–4 cells wide and 10–20 cells tall (Fig. D) (up to 40 in *A. excelsa*, Fig. C). Rays heterocellular, typically with one row of square and/or upright marginal cells (Fig. F) (also homocellular in *A. indica*). Prismatic crystals present in the axial parenchyma in chains of chambered cells (Fig. G). Tangential lines of traumatic resin canals occasionally present (Fig. B).

The wood anatomy of *Azadirachta* is described in Kribs (1930), Pearson and Brown (1932), Metcalfe and Chalk (1950), Ghosh *et al.* (1963), Pennington and Styles (1975), Wong (1976), Datta and Samanta (1983), Nair (1988), Lemmens *et al.* (1995), Negi *et al.* (2003), Shah (2004), and the InsideWood website (2004 onwards).

## Diagnostic characteristics

There is little distinction to be made between *A. excelsa* and *A. indica*. Slight differences occur in ray height and cellular composition, but these are not enough to tell the woods apart.

Both species of *Azadirachta* may be distinguished from *Swietenia*. The two genera have similar banded and scanty to vasicentric parenchyma, and the intervessel pitting and rays are very similar. However, the abundance of prismatic crystal chains seen in the axial parenchyma of *Azadirachta* contrasts with the individual crystals found in the ray cells of *Swietenia*. As in *Melia* (Fig. E, p. 61), there are striations on the vessel walls that do not occur in *Swietenia* or any of the other mahogany woods.

Although it can be very difficult to distinguish between the wood anatomy of *Melia composita* and *Azadirachta*, the distinctive vessel pattern in *M. azedarach* (Fig. B, p. 61) is not seen in *Azadirachta*, and banded parenchyma is far more abundant in *Azadirachta*.

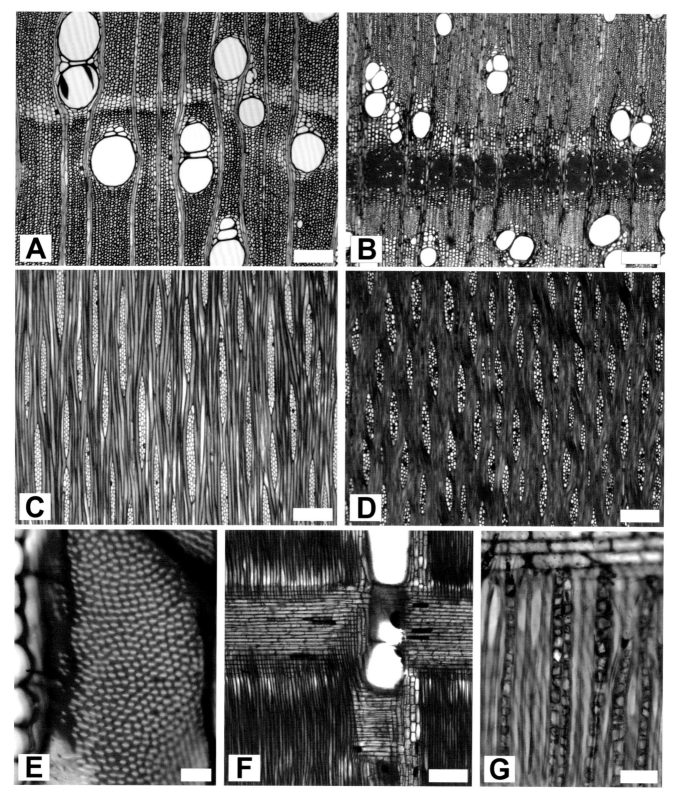

**Figures A, C, E and F** *Azadirachta excelsa*, **Figures B, D and G** *Azadirachta indica*. **A and B.** TS, wood diffuse porous, paratracheal axial parenchyma scanty and vasicentric, apotracheal axial parenchyma in bands 3–6 cells wide. Note also the traumatic resin canals in Fig. B. **C and D.** TLS, rays 2–4 cells wide and up to 40 cells high in *A. excelsa* (Fig. C), and up to 20 cells high in *A. indica* (Fig. D). **E.** TLS, intervessel pitting alternate and minute to small. **F.** RLS, rays heterocellular. **G.** RLS, abundant chains of prismatic crystals in chambered axial parenchyma cells. Figs. A–D and F, scale bar = 200 μm, Fig. E, scale bar = 20 μm, Fig. G, scale bar = 50 μm.

# *Cabralea canjerana* (Vell.) Mart.

## Introduction

*Cabralea* is a monotypic genus, and *C. canjerana* is an important construction timber in tropical America, in particular Brazil. *Cabralea canjerana* shares similarities to, and is often confused with, *Cedrela* (pp. 26–29), as they have overlapping geographic distributions and share some morphological characteristics.

## Natural distribution

*Cabralea canjerana* extends from Costa Rica to the Guianas, Bolivia, Peru, Brazil, northern Argentina, Paraguay and Uruguay.

Countries in which ***Cabralea canjerana*** occurs naturally

## Wood

The timber has a pink to dark red or maroon colouring.

## Uses

*Cabralea* is used in the same way as other mahogany timbers, in the production of interiors, furniture, cabinets, and in house construction. The timber is used extensively for construction in Brazil, and also has applications in cabinet work, boarding, and as a dye (from the sawdust) and medicinal resource (from the bark).

The timber is often used locally as a substitute for *Cedrela* (p. 27) due to similarities in wood properties.

## Common names

There are many similar names in use for *C. canjerana*. These include Canjerana, Canyarana, Cancharana, Caroba, Requia blanca, Cedro masha and Cedrahy. The last two names are similar to those for *Cedrela*, which is often confused with *Cabralea*. In Brazil the common name Pau de Santo is used, because images of saints were carved from *Cabralea* wood.

***Cabralea canjerana*** transverse surface, × 5

***Cabralea canjerana*** longitudinal surface, × 1 (inset) and × 1.5

## Anatomy

Figures A–D

Growth ring boundaries indistinct or absent. Wood diffuse-porous. Vessels solitary and in radial multiples of up to four (Fig. A). Simple perforation plates. Intervessel pits small and alternate (Fig. C). Vessel-ray pits with distinct borders and similar in size and shape to intervessel pits. Fibre walls thin to thick. Septate fibres present (Fig. B). Paratracheal parenchyma aliform and confluent, forming bands 3–5 cells wide (Fig. A), occasionally up to 8 cells wide. Rays not storied, mainly uniseriate, less often biseriate, and 5–25 cells high (Fig. B). Rays heterocellular with one row of square and/or upright marginal cells (Fig. D). No mineral inclusions seen.

References to the wood anatomy of *Cabralea canjerana* include Moeller (1876), Record (1924), Kribs (1930), Metcalfe and Chalk (1950), Kribs (1959), Brazier and Franklin (1961), Pennington and Styles (1975), Détienne and Jacquet (1983), Acevedo Mallque and Kikata (1994), Barros and Callado (1997), and the InsideWood website (2004 onwards).

## Diagnostic characteristics

The wood of *Cabralea canjerana* is different to that of *Swietenia*. The banded parenchyma in *Cabralea canjerana* is associated with the vessels (confluent), in contrast to the apotracheal banded parenchyma in *Swietenia*. The intervessel pits in *Cabralea canjerana* are larger than the minute pits of *Swietenia* (and many other mahogany woods), and the rays are unstoried and narrow with no crystal inclusions, compared to the wide storied rays of *Swietenia* that often include prismatic crystals.

Although the timbers of *Cabralea canjerana* and *Cedrela* can be confused, the wood anatomy is different. *Cedrela* is easily distinguished from *Cabralea canjerana* as it has semi-ring or ring-porous wood, and prismatic crystals (pp. 28–29).

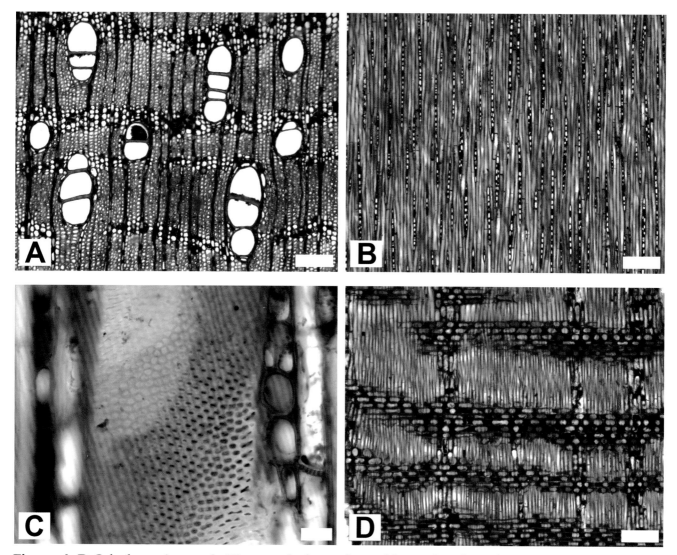

**Figures A–D** *Cabralea canjerana.* **A.** TS, paratracheal parenchyma aliform and confluent, forming indistinct bands 3–5 cells wide. Vessels solitary and in radial multiples of 2–4. **B.** TLS, rays uniseriate and biseriate, and of varying heights (usually 5–25 cells tall), fibres septate. **C.** TLS, alternate and small intervessel pits. **D.** RLS, rays heterocellular. Figs. A, B, D, scale bar = 200 μm, Fig. C, scale bar = 20 μm.

# *Carapa* Aubl.

## Introduction

The genus *Carapa* consists of three or four species from tropical America, the Caribbean, and west and central Africa. The wood is considered to be less attractive than *Swietenia* but has similar properties. Of these species, *C. guianensis* and *C. procera* are most commonly used and traded internationally. *Carapa* is closely related to *Xylocarpus*, a mangrove tree that is not an internationally traded timber.

## Natural distribution

Both *C. guianensis* and *C. procera* have been introduced into South–East Asia (mainly Java, Peninsular Malaysia and Singapore) where they may have potential as plantation species. However, their natural distribution is as follows:

***Carapa guianensis*** is widespread, occuring in Central America (Guatemala to Panama) and South America (Colombia, Ecuador, Peru, Brazil, Venezuela and Suriname) and the Caribbean.

***Carapa procera*** is restricted to Brazil, Suriname and French Guiana in South America, and also occurs in west and central Africa.

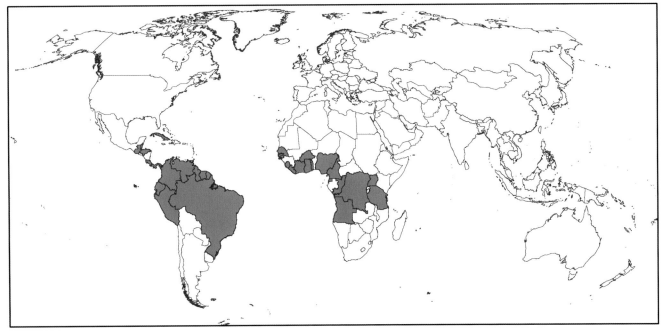

Countries in which ***Carapa guianensis*** and ***Carapa procera*** occur naturally

# *C. guianensis* Aubl. and *C. procera* DC.

## Wood

The colour of *Carapa* timber varies from yellow brown to pink/red brown, resembling the variation in *Swietenia*.

## Uses

The timber of both *C. guianensis* and *C. procera* is heavily utilised locally and is exported. Its uses are similar to those of other mahoganies, as it is used in furniture production (cabinets particularly), flooring and veneer. The bark has medicinal properties and the seed oil is used in soap production.

## Common names

There are various names common to both species of *Carapa* that are used locally and internationally. *Carapa* in general is commonly known as Crabwood, with many common names and more specific names attributed to different species.

*Carapa guianensis* is known as Crappo, Andiroba, Caobilla and Tangare. *Carapa procera* is known as Andiroba, as well as Krappa and African crabwood.

***Carapa guianensis*** transverse surface, × 4.5

***Carapa procera*** longitudinal surface, × 1 (inset) and × 1.25

## Anatomy

Figures A–H

Growth rings present but indistinct. Wood diffuse-porous. Solitary vessels common (Fig. A), also short radial multiples and clusters of 2–3 (Fig. B). Simple perforation plates. Intervessel pits alternate, minute to small in *C. guianensis* (Fig. D), and minute in *C. procera* (Fig. E). Vessel-ray pits with distinct borders and similar in size and shape to intervessel pits. Fibre walls thin to thick. Septate fibres present (Fig. C). Apotracheal parenchyma in bands, 1–3 cells wide. Paratracheal parenchyma scanty and vasicentric (Fig. A), and also lozenge to winged-aliform in *C. procera* (Fig. B). Heterocellular rays with several rows of square and/or upright marginal cells, often up to 4 (Fig. F). Rays not storied in *C. guianensis* (Fig. C), irregularly storied in *C. procera*. Rays 1–3 cells wide, and 6–30 cells high. Prismatic crystals present in the ray cells (Fig. G), and in chambered cells in the axial parenchyma (Fig. H).

Wood anatomy references to *Carapa* include Moeller (1876), Dixon (1919), Koehler (1922), Record (1924), Kribs (1930), Pearson and Brown (1932), Rendle (1938), Lebacq and Istas (1950), Metcalfe and Chalk (1950), Fouarge *et al.* (1953), Normand (1955), Kribs (1959), Brazier and Franklin (1961), Lebacq (1963), Willemstein (1975), Normand and Paquis (1976), Gaiotti de Peralta and Edlmann Abbate (1981), Détienne *et al.* (1982), Datta and Samanta (1983), Détienne and Jacquet (1983), Freitas (1987), Acevedo Mallque and Kikata (1994), Sosef *et al.* (1998), and the InsideWood website (2004 onwards).

## Diagnostic characteristics

*Carapa guianensis* and *C. procera* are identifiable by their paratracheal parenchyma, as both species have scanty and vasicentric parenchyma, but in *C. procera* this extends to lozenge and winged-aliform. The anatomy of *Carapa* is quite different from that of *Swietenia*. The banded apotracheal parenchyma in *Carapa* is less well-defined than it can be in *Swietenia*, and the vasicentric and aliform parenchyma (in *C. procera*) is clearly different from the scanty paratracheal parenchyma of *Swietenia*.

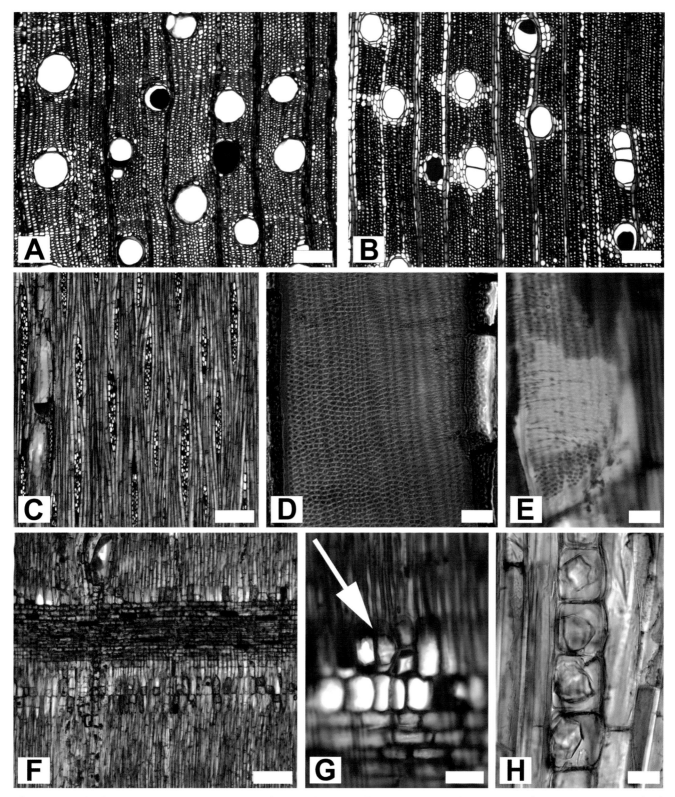

**Figures A, C, D, F, H *Carapa guianensis*, Figures B, E, G *Carapa procera*. A.** TS, parenchyma scanty paratracheal and vasicentric. **B.** TS, vasicentric and lozenge-aliform paratracheal parenchyma. **C.** TLS, rays not storied, up to 3 cells wide, and up to 30 cells in height. Fibres septate. **D.** TLS, intervessel pitting minute to small and alternate. **E.** TLS, minute and alternate intervessel pitting. **F.** RLS, heterocellular ray, with several rows of square and/or upright marginal cells. **G.** RLS, prismatic crystals in upright ray cells (see arrow). **H.** RLS, chambered axial parenchyma cells with individual prismatic crystals. Figs. A–C, F, scale bar = 200 μm, Figs. D–E, scale bar = 20 μm, Fig. G, scale bar = 50 μm.

# *Cedrela* P.Browne

## Introduction

*Cedrela* is a genus from tropical America that includes 17 species. *Cedrela odorata* and *C. fissilis* are commonly traded internationally and hence are featured in detail here. *Cedrela odorata* is an important timber, listed on CITES Appendix III, meaning that its trade is monitored and is subject to permit.

## Natural distribution

*Cedrela* is widespread across Central and South America (excluding Chile). It has also been established as a plantation tree in Africa (Uganda, Tanzania, South Africa) and South-East Asia (Thailand, Peninsular Malaysia, Singapore and the Philippines). The distribution of the two species covered here is as follows:

**Cedrela fissilis** has a wide distribution from Panama to northern Argentina.

**Cedrela odorata** is found in Mexico through Central America to northern Argentina, and also in the Caribbean.

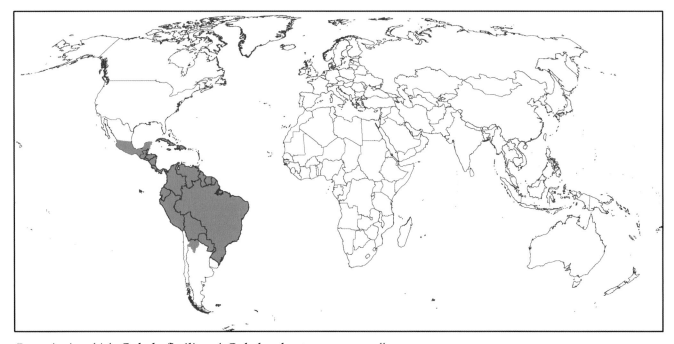

Countries in which **Cedrela fissilis** and **Cedrela odorata** occur naturally

# *C. fissilis* Vell. and *C. odorata* L.

## Wood

The timber of *Cedrela* has a pale golden brown to pink/brown colour, similar to that of *Swietenia*, and may have a cedar-like odour. Although *Cedrela* shares similarities with *Swietenia*, it is used as a timber in its own right, rather than as a mahogany substitute.

## Uses

*Cedrela* has traditionally been used in the production of cigar boxes, but it is also used in a similar manner to other mahogany woods, for example in construction, furniture, and as a veneer.

## Common names

*Cedrela* is commonly known as Cedar, whereas true cedar is a softwood (*Cedrus* species). Other general names for *Cedrela* include Red cedar (this name is also used for *Toona* (p. 71), Spanish cedar, Cigarbox cedar and Stinking mahogany.

More specifically, *C. fissilis* is known as South American cedar, Brazilian cedar or Cedro (Cedro batata, Cedro blanco, Cedro colorado, Cedro rosado). *Cedrela odorata* is widely known as Cedro in many Latin American countries, and many other names based on location, such as Central American cedar, Honduras cedar, Tabasco cedar, Acaju, Acajou rouge, Acajou-bois and Cedrat.

***Cedrela fissilis*** transverse surface, × 5

***Cedrela odorata*** longitudinal surface, × 1

## Anatomy

Figures A–F

Growth rings distinct, wood ring-porous in *C. odorata* (Fig. A), semi-ring-porous in *C. fissilis* (Fig. B). Vessels mainly solitary. Perforation plates simple. Intervessel pits small to medium, alternate (Fig. D). Vessel-ray pits with distinct borders and similar in size and shape to intervessel pits. Fibre walls thin to thick. Fibres mostly non-septate, but some septate fibres occasionally present. Apotracheal parenchyma diffuse, initial (Fig. A) and indistinct initial (Fig. B), forming marginal bands up to 8 cells wide, paratracheal parenchyma scanty and vasicentric to aliform (Figs. A and B). Rays heterocellular with one row (occasionally up to four rows) of square and/or upright marginal cells (Fig. E). Rays 2–3 cells wide and up to 20 cells high in *C. odorata* (Fig. C), and up to 30 in *C. fissilis*, not storied. Prismatic crystals present in ray and non-chambered axial parenchyma cells (Fig. F).

References to the wood anatomy of *Cedrela* include Moeller (1876), Dixon (1919), Koehler (1922), Record (1924), Kribs (1930), Pearson and Brown (1932), Rendle (1938), Metcalfe and Chalk (1950), Kribs (1959), Brazier and Franklin (1961), Ghosh *et al.* (1963), Détienne *et al.* (1982), Détienne and Jacquet (1983), Donaldson (1984), Espinoza de Pernia (1987), Acevedo Mallque and Kikata (1994), Lemmens *et al.* (1995), Dünisch, Bauch and Gasparotto (2002), and the InsideWood website (2004).

## Diagnostic characteristics

It is difficult to tell apart the individual species of *Cedrela*, but *C. odorata* is ring-porous compared to the semi-ring-porous *C. fissilis*, and the parenchyma in *C. fissilis* tends to be more aliform than in *C. odorata*.

The ring-porous or semi-ring-porous wood of *Cedrela* is strikingly different from the diffuse-porous wood of *Swietenia* and most other mahoganies. The only other Meliaceae mahoganies with similar wood porosity are *Melia* (pp. 60–61) and *Toona* (pp. 72–73) from India and South-East Asia. In addition, the paratracheal parenchyma in *Cedrela* has a tendency to be aliform (especially in *C. fissilis*), and the rays are smaller and unstoried in *Cedrela*, compared to *Swietenia*.

The wood anatomy of *Cedrela* greatly resembles that of *Toona* (pp. 72–73), and the two are easily confused unless the geographical origin is known.

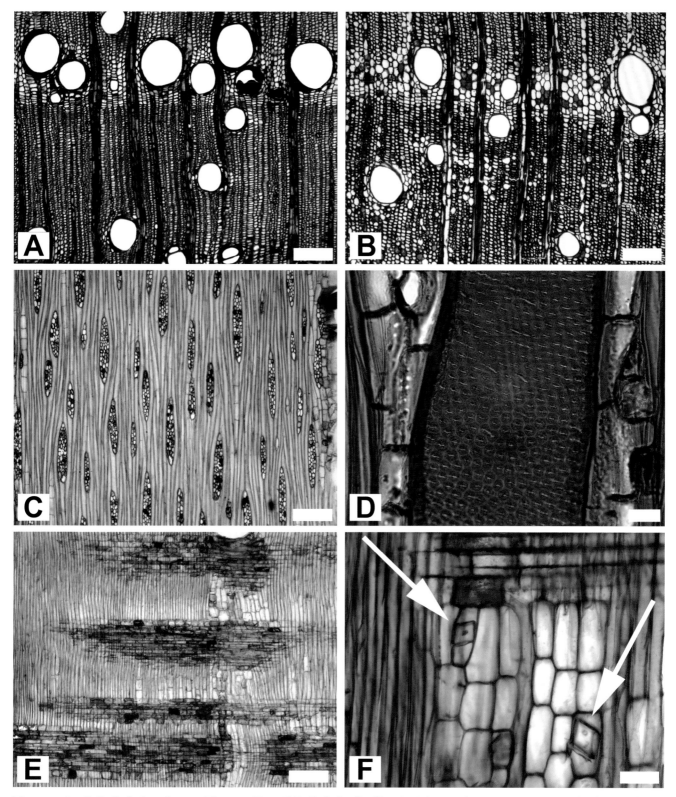

**Figures A, C–E** *Cedrela odorata*, **Figures B and F** *Cedrela fissilis*. **A.** TS, wood ring-porous, apotracheal parenchyma forming a marginal (initial) band, and paratracheal parenchyma vasicentric (to aliform). **B.** TS, wood semi ring-porous, forming an indistinct initial band of apotracheal parenchyma, and paratracheal parenchyma scanty, vasicentric and aliform. **C.** TLS, rays not storied. **D.** TLS, intervessel pits small to medium and alternate. **E.** RLS, heterocellular rays with one row of square and/or upright marginal cells. **F.** RLS, prismatic crystals present in non-chambered axial parenchyma cells (see arrows). Figs. A, B, C, E, scale bar = 200 μm, Fig. D, scale bar = 20 μm, Fig. F, scale bar = 50 μm.

# *Chisocheton divergens* Blume

## Introduction

There are over 50 species of *Chisocheton*, in India, South-East Asia and Australia. The timber is not of major commercial importance, but is similar in appearance to *Swietenia*, and is often confused anatomically with *Guarea*, and therefore warrants inclusion here. As *Chisocheton* is such a large genus, *C. divergens* has been chosen here as a representative, based on its similarity in appearance to other mahogany woods. *Chisocheton divergens* may also be known by the synonym *C. patens*.

## Natural distribution

*Chisocheton divergens* is widely distributed across China, Singapore, Malaysia, Indonesia and Taiwan.

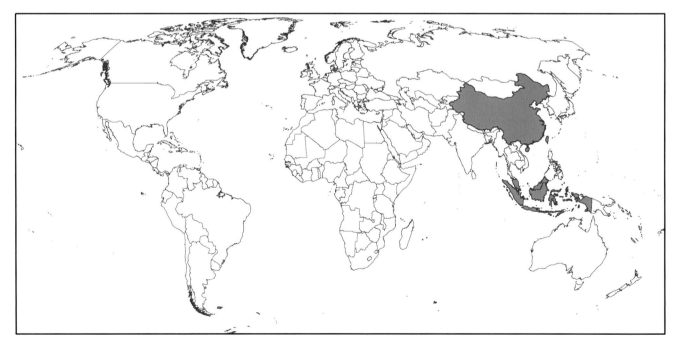

Countries in which *Chisocheton divergens* occurs naturally

## Wood

The wood ranges from yellow to pale brown or pale pink, and is not as dark as the majority of Meliaceae mahoganies.

## Uses

The timber is widely used in the production of furniture, in construction and ship building, as a veneer and as a pulp. Other parts of the tree that are used include oil from the seeds, and edible fruits.

## Common names

*Chisocheton* is known by local names such as Chisocheton, Lantupak and Kiso.

***Chisocheton divergens*** transverse surface, × 4

***Chisocheton divergens*** longitudinal surface, × 1 (inset) and × 2.5

## Anatomy

Figures A–E

Growth rings distinct, wood diffuse-porous. Vessels solitary and in radial multiples of up to 4. Perforation plates simple. Intervessel pits minute to small, alternate (Fig. C). Vessel-ray pits with distinct borders and similar in size and shape to intervessel pits. Fibre walls thin to thick. Septate fibres present (Fig. B). Paratracheal parenchyma reticulate, forming bands commonly 2–3 cells wide (Fig. A) and occasionally marginal. Rays heterocellular with one row (occasionally up to four rows) of square and/or upright marginal cells and not storied (Fig. D). Rays 1–3 cells wide and up to 20 cells high (Fig. B). Prismatic crystals present in short chains of chambered axial parenchyma. Silica bodies present in ray cells (Fig. E), axial parenchyma and fibres.

References for *Chisocheton* include Metcalfe and Chalk (1950), Ghosh *et al.* (1963), Pennington and Styles (1975), Datta and Samanta (1983), Sosef *et al.* (1998), Negi *et al.* (2003), and the InsideWood website (2004 onwards).

## Diagnostic characteristics

The banded reticulate parenchyma in *C. divergens* is distinctly different to the banded apotracheal parenchyma in *Swietenia*. When the bands of parenchyma in *C. divergens* are wider than 4 cells, there is a close similarity with *Guarea* (pp. 48–49) and *Dysoxylum* (pp. 40–41).

The other characteristics that differentiate *C. divergens* from *Swietenia* include narrower rays, larger intervessel pitting and the presence of silica bodies (also present in *Entandrophragma candollei* (p. 44), *Guarea* (p. 48), and *Aphanamixis polystachya* (p. 12).

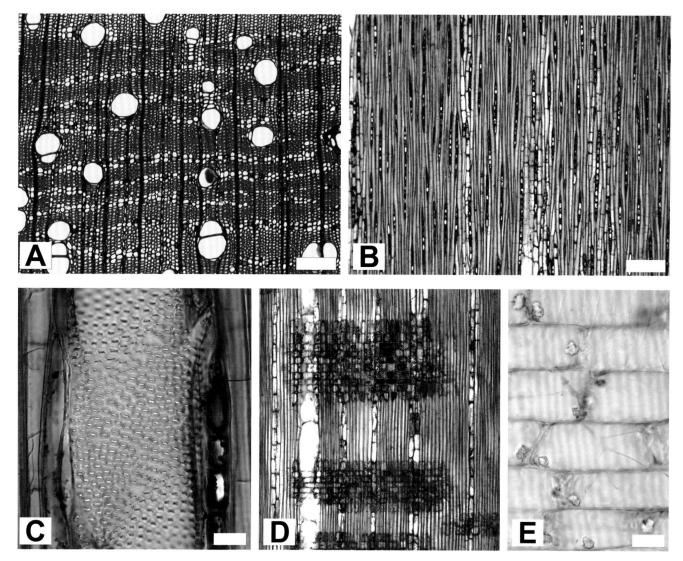

**Figures A–E** *Chisocheton divergens*. **A.** TS, paratracheal parenchyma reticulate, forming bands 2–3 cells wide, occasionally marginal. **B.** TLS, rays narrow and unstoried, abundant septate fibres. **C.** TLS, small, alternate intervessel pitting. **D.** RLS, rays heterocellular. **E.** RLS, silica bodies in procumbent ray cells. Figs. A, B, D, scale bar = 200 μm, Figs. C and E, scale bar = 20 μm.

# *Chukrasia tabularis* A.Juss.

## Introduction

*Chukrasia tabularis* is a major commercial timber and is one of possibly two species that belong to the genus. It is one of the major mahogany substitutes from India, along with *Melia* (pp. 58–61), *Azadirachta* (pp. 14–17), and *Aphanamixis polystachya* (pp. 10–13).

## Natural distribution

The natural distribution of *C. tabularis* is mainly in India, Bangladesh, Thailand (west) and Peninsular Malaysia (north). It also occurs in Pakistan, Sri Lanka, southern China, Sumatra and Borneo. Plantations exist in Africa and tropical America.

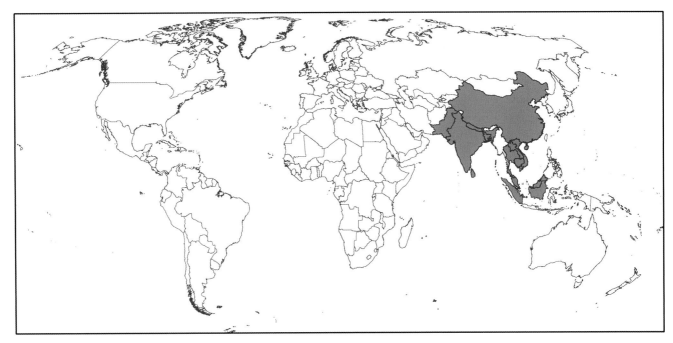

Countries in which ***Chukrasia tabularis*** occurs naturally

## Wood

The wood of *C. tabularis* is characteristically red/brown, but can vary between more yellow or darker brown.

## Uses

As with many mahogany woods, *C. tabularis* has an attractive finish and is hence used in furniture production (cabinet making) and decorative panelling. It also has applications in ship and boat building, construction work and as a veneer.

## Common names

*Chukrasia tabularis* is known by a variety of names based upon country of origin, language or trade. Commonly used names include Chickrassy, Yinma, Chittagong wood, Yom-hin, Repoh, Surian batu, Cherana puteh and Suntang.

*Chukrasia tabularis* transverse surface, × 4

*Chukrasia tabularis* longitudinal surface, × 1

## Anatomy

Figures A–E

Growth rings indistinct. Wood diffuse-porous. Vessels solitary, also in radial multiples of 2–3 (Fig. A). Perforation plates simple. Intervessel pitting alternate and minute to small (Fig. C). Vessel-ray pits with distinct borders and similar in size and shape to intervessel pits. Fibre walls thin to thick. All fibres non-septate. Apotracheal axial parenchyma banded and marginal (commonly 3 cells wide) (Fig. A). Paratracheal axial parenchyma scanty (occasionally vasicentric) (Fig. A). Rays not storied, mainly uniseriate and biseriate, also up to 3 cells wide (Fig. B) and up to 15 cells high. Rays heterocellular with one row of square and/or upright marginal cells (Fig. D). Prismatic crystals abundant, present in the axial parenchyma, in chambered cells, and often in chains (Fig. E).

References to the wood anatomy of *Chukrasia* include Kribs (1930), Pearson and Brown (1932), Panshin (1933), Metcalfe and Chalk (1950), Desch (1954), Kribs (1959), Brazier and Franklin (1961), Ghosh *et al.* (1963), Wong (1976), Datta and Samanta (1983), Lemmens *et al.* (1995), Negi *et al.* (2003), and the InsideWood website (2004 onwards).

## Diagnostic characteristics

Several differences separate the wood anatomy of *C. tabularis* and *Swietenia*. Both woods have banded apotracheal parenchyma and scanty paratracheal parenchyma, but the parenchyma is more vasicentric in *C. tabularis*. In addition, the intervessel pitting in *C. tabularis* is minute to small, the rays are not storied, and there are abundant prismatic crystals in chains, all features not present in *Swietenia*.

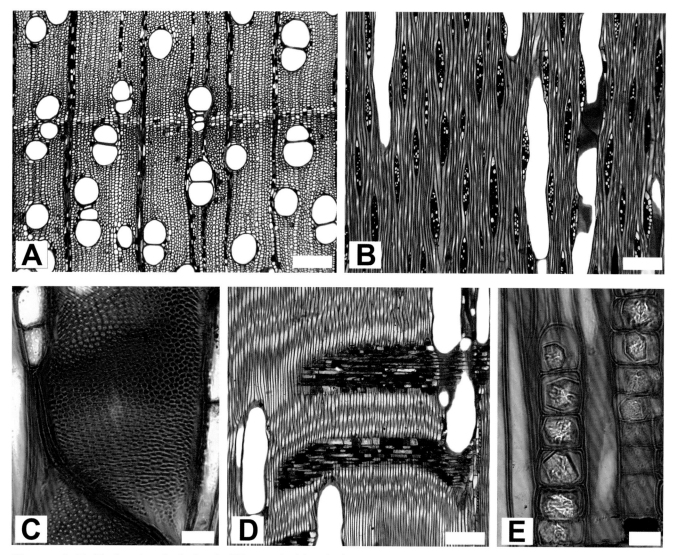

**Figures A–E *Chukrasia tabularis.*  A.** TS, marginal band of apotracheal parenchyma, paratracheal parenchyma scanty (to vasicentric).  **B.** TLS, rays not storied.  **C.** TLS, minute to small intervessel pitting.  **D.** RLS, rays heterocellular.  **E.** RLS, chains of prismatic crystals in chambered axial parenchyma cells. Figs. A, B, D, scale bar = 200 μm, Figs. C and E, scale bar = 20 μm.

# *Dysoxylum fraserianum* Benth.

## Introduction

The genus *Dysoxylum* includes approximately 80 species. *Dysoxylum fraserianum* is internationally traded and often known as a mahogany timber.

## Natural distribution

*Dysoxylum* is widespread from India and Sri Lanka to South-East Asia, including south China, and to Australasia, with one species in New Zealand (North Island). *Dysoxylum fraserianum* is one of three *Dysoxylum* species native to Australia, and is found in New South Wales and Queensland.

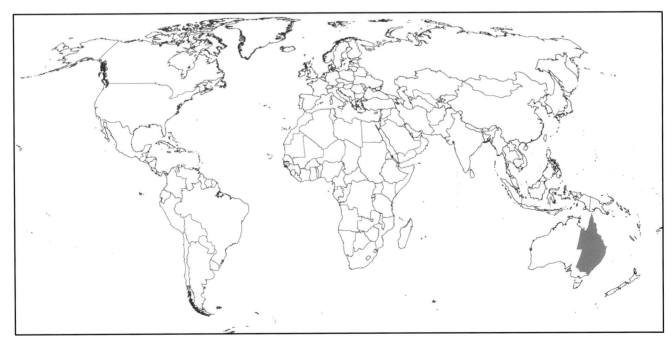

Countries in which **Dysoxylum fraserianum** occurs naturally

## Wood

The wood of *D. fraserianum* ranges in colour, from light yellow/brown to dark brown/red, often resembling *Swietenia* and other mahogany woods.

## Uses

Due to its colour *Dysoxylum* is used in cabinet making, as well as decking and furniture. Species of *Dysoxylum* that are less attractive and do not resemble true mahogany timber are used locally, mostly in construction, rather than traded internationally.

## Common names

There are many names associated with *Dysoxylum*, including Dysox, Membalun and Cempaga. *Dysoxylum fraserianum* in particular is known as Australian mahogany, Australian rosewood, Rose mahogany or Rosewood.

***Dysoxylum fraserianum*** transverse surface, × 4

***Dysoxylum fraserianum*** longitudinal surface, × 1

## Anatomy

Figures A–E

Growth rings absent. Wood diffuse-porous. Vessels solitary and in radial multiples of 2–3 (Fig. A). Perforation plates simple. Intervessel pits alternate and variable in size, from small to medium (Fig. C). Vessel-ray pits with distinct borders and similar in size and shape to intervessel pits. Fibre walls thick to very thick. Septate fibres present (Fig. B). Paratracheal axial parenchyma confluent, forming irregular bands, 3–4 cells wide (Fig. A), and occasionally vasicentric. Rays not storied, mainly biseriate (also 1–3 cells wide), and typically between 10 and 20 cells high (Fig. B). Rays heterocellular with one row of square and/or upright marginal cells (Fig. D). Prismatic crystals abundant, present in the axial parenchyma, in chambered cells and often in chains (Fig. E).

References to the wood anatomy of *Dysoxylum* include Kribs (1930), Pearson and Brown (1932), Metcalfe and Chalk (1950), Desch (1954), Brazier and Franklin (1961), Ghosh *et al.* (1963), Pennington and Styles (1975), Datta and Samanta (1983), Sosef *et al.* (1998), and the InsideWood website (2004, onwards).

## Diagnostic characteristics

The wood of *D. fraserianum* is quite distinct from that of *Swietenia* and the other Meliaceae mahoganies. One of the most striking characteristics of *D. fraserianum* is the parenchyma that forms wide wavy bands, which are similar to those found in *Entandrophragma* (pp. 44–45) and *Guarea* (pp. 48–49), and quite distinct from the banded apotracheal parenchyma in *Swietenia*. In addition, the rays are narrow and unstoried, there are abundant crystals, and the intervessel pits are clearly larger than the minute pits in *Swietenia*.

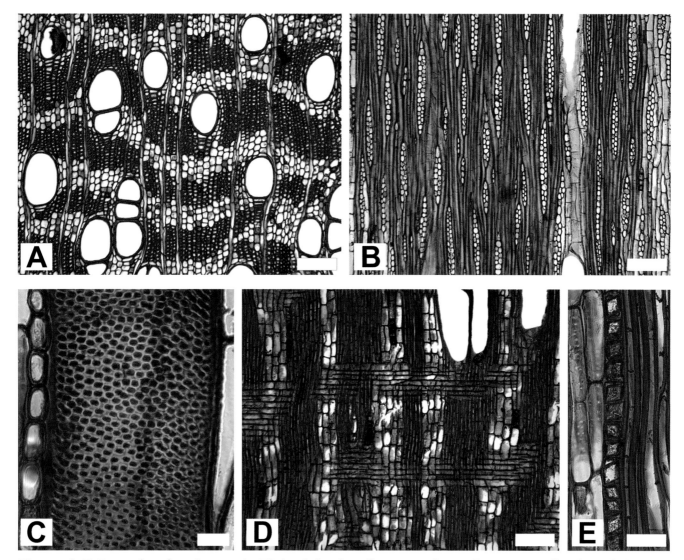

**Figures A–E** *Dysoxylum fraserianum.* **A.** TS, banded and confluent parenchyma. **B.** TLS, rays tall and mainly biseriate, septate fibres. **C.** TLS, intervessel pits alternate and small to medium. **D.** RLS, rays heterocellular. **E.** RLS, prismatic crystals in chambered axial parenchyma cells. Figs. A, B, D, scale bar = 200 μm, Fig. C, scale bar = 20 μm, Fig. E, scale bar = 50 μm.

# *Entandrophragma* C.DC.

## Introduction

*Entandrophragma* consists of 11 species, not all of which are commercially important. *Entandrophragma cylindricum* is one of the most abundant and commercially valuable species, and *E. angolense*, *E. candollei* and *E. utile* are also internationally traded commercial timbers. The other seven species are less significant in international trade.

## Natural distribution

*Entandrophragma* extends across west, central and east Africa. The distribution of the four species covered here is as follows:

***Entandrophragma angolense*** occurs across west and east Africa, from Angola and the Ivory Coast to Uganda and Sudan.

***Entandrophragma candollei*** is found in west and central Africa, from Guinea and the Ivory Coast, south to Angola and Zaire.

***Entandrophragma cylindricum*** extends across central Africa, from the west in Ghana, Nigeria and Cameroon, and to the east in Zaire, Tanzania and Uganda.

***Entandrophragma utile*** occurs in tropical west Africa in Sierra Leone, Liberia, Cameroon, Gabon and east to Uganda.

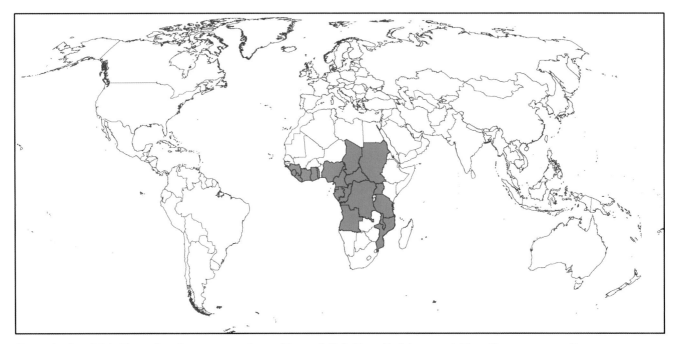

Countries in which ***Entandrophragma angolense***, ***E. candollei***, ***E. cylindricum*** and ***E. utile*** occur naturally

# E. *angolense* (Welw.) C.DC., E. *candollei* Harms, E. *cylindricum* Sprague and E. *utile* Sprague

## Wood

*Entandrophragma* timber is similar to the red/brown appearance of *Swietenia*, with some variation in the darkness of the wood. *Entandrophragma cylindricum* often has a strong scent when freshly cut.

## Uses

*Entandrophragma* is a high-quality timber used as a mahogany substitute, especially in the production of furniture, cabinets and veneers. The wood is used locally for canoe and boat building.

## Common names

Sapele mahogany is broadly used as a name for *Entandrophragma*, but commonly associated with *E. cylindricum*. There are various trade and local names associated with different species. For example, *E. angolense* is commonly known as Tiama, Gedu nohor, Edinam and Penkwa. *Entandrophragma candollei* is known as Omu, Kosipo and Heavy sapele. *Entandrophragma cylindricum* is referred to as Aboudikro and Scented mahogany. *Entandrophragma utile* is known as Sipo, Utile and Assie.

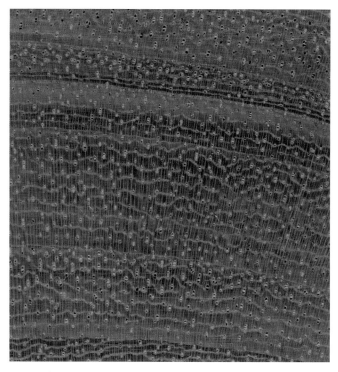

***Entandrophragma utile*** transverse surface, × 4

***Entandrophragma utile*** longitudinal surface, × 1 (inset) and × 1.5

## Anatomy

Figures A–G

Growth rings indistinct, wood diffuse-porous. Vessels commonly solitary and in radial pairs (Figs. A and B), also in clusters (particularly in *E. utile*). Perforation plates simple. Alternate and minute intervessel pits (Fig. D). Vessel-ray pits with distinct borders and similar in size and shape to intervessel pits. Fibre walls thin to very thick. Fibres mostly non-septate, but septate fibres may be present. Paratracheal parenchyma scanty and vasicentric (Figs. A and B), occasionally extending to resemble lozenge-aliform. Paratracheal parenchyma also in irregular marginal bands 2–5 cells wide (Fig. A). Rays irregularly storied, often in echelon (Fig. C). Rays heterocellular with one row of square and/or upright marginal cells (Figs. C and E), 3–5 cells wide, up to 20 cells high. Prismatic crystals often present in the marginal ray cells (Fig. F) and in chains of chambered cells in the axial parenchyma (Fig. G). Silica bodies present in ray cells and axial parenchyma in *E. candollei*. Tangential lines of traumatic resin canals sometimes present (Fig. A).

The wood anatomical references to *Entandrophragma* include Dixon (1919), Kribs (1930), Panshin (1933), Rendle (1938), Eggeling and Harris (1939), Lebacq and Istas (1950), Metcalfe and Chalk (1950), Fouarge *et al.* (1953), Normand (1955), Kribs (1959), Brazier and Franklin (1961), Lebacq (1963), Wagenführ and Steiger (1963), Normand and Paquis (1976), and the InsideWood website (2004 onwards).

## Diagnostic characteristics

The *Entandrophragma* species detailed here have almost identical wood anatomy. As with other woods, variation occurs between samples of the same species, making the distinction between individual species difficult. However, the marginal bands of parenchyma in *E. cylindricum* and *E. utile* are not present in *E. angolense*. *Entandrophragma candollei* is distinguishable from the other species by the presence of silica bodies in the rays and axial parenchyma. There is a detailed comparison of *Entandrophragma* species in Brazier and Franklin (1961).

The wood of *Entandrophragma* is distinct from that of *Swietenia* and the other mahogany woods. The characteristics that distinguish *Entandrophragma* are the banded paratracheal parenchyma seen in *E. cylindricum* and *E. utile*, and the occurrence of storied rays in echelon. The chains of crystals in chambered axial parenchyma cells that often occur in *Entandrophragma* (*E. cylindricum* in particular) are a distinctive feature when present. Of the Meliaceae mahogany timbers, silica bodies are seen only in *E. candollei*, *Guarea* (p. 48), *Aphanamixis polystachya* (p. 12), and *Chisocheton divergens* (pp. 32–33). Although bands of traumatic resin canals do occur in *Entandrophragma*, they are often absent, and therefore not diagnostic.

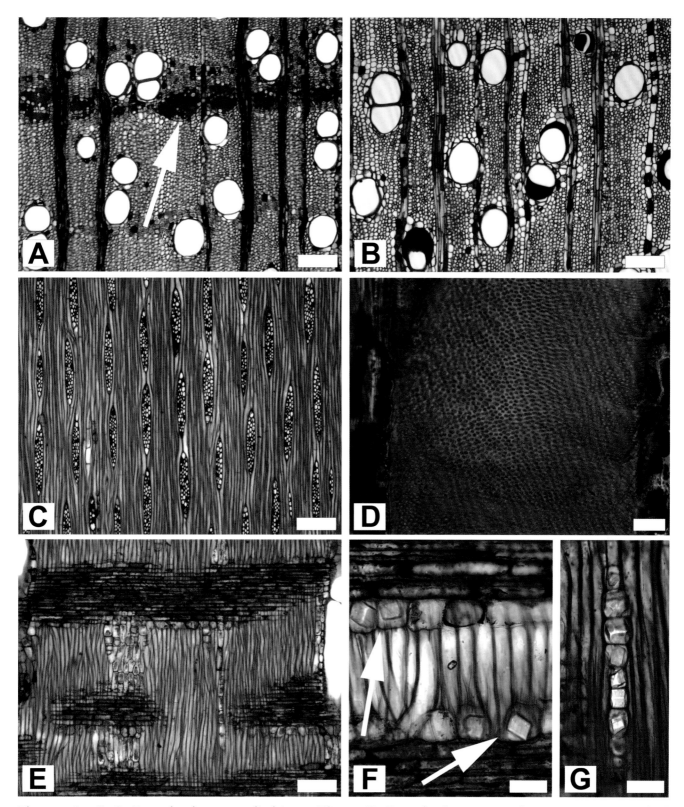

**Figures A, C–G *Entandrophragma cylindricum*, Figure B *Entandrophragma angolense*. A.** TS, paratracheal parenchyma vasicentric, irregular marginal bands of parenchyma, 2–5 cells wide. Tangential band of traumatic resin canals (see arrow). **B.** TS, vasicentric paratracheal parenchyma. **C.** TLS, storied rays in echelon and mostly non-septate fibres. **D.** TLS, minute, alternate intervessel pitting. **E.** RLS, rays heterocellular. **F and G.** RLS, prismatic crystals in square marginal ray cells (see arrows) and in chains of chambered axial parenchyma cells. Figs. A, B, C, E, scale bar = 200 μm, Fig. D, scale bar = 20 μm, Figs. F and G, scale bar = 50 μm.

# *Guarea* L.

## Introduction

There are approximately 50 species of *Guarea*, spread mainly across tropical America and also in tropical Africa. The American species are generally small trees or shrubs, such as G. *guidonia,* which is from Central and South America and the Caribbean, and G. *kunthiana* which is widespread in South America. *Guarea glabra* is a large tree that is used locally for construction in South America.

International trade in *Guarea* timber is primarily in the African species, which include G. *cedrata* and G. *thompsonii. Guarea cedrata* is of greater commercial importance.

## Natural distribution

Both species are found in west tropical Africa, from Sierra Leone, Nigeria and the Ivory Coast to the Congo.

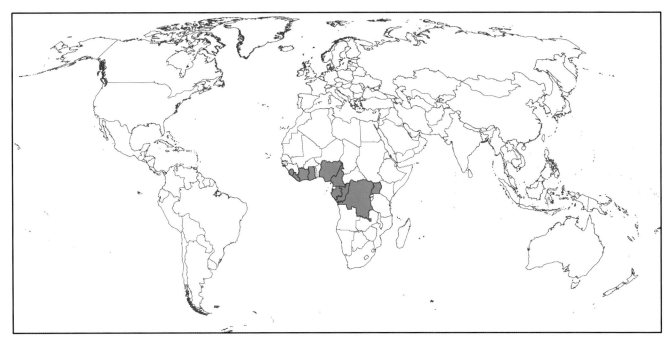

Countries in which *Guarea cedrata* and *Guarea thompsonii* occur naturally

# *G. cedrata* Pellegr. ex A.Chev. and *G. thompsonii* Sprague & Hutch.

## Wood

The colour of *Guarea* timber resembles that of *Swietenia* and other mahogany woods, and can vary between brown and red/brown in both *G. cedrata* and *G. thompsonii*.

## Uses

*Guarea* timber is used in construction, furniture production, boat building, and as a decorative veneer.

## Common names

Both species are known as African cedar, Scented mahogany, or Cedar mahogany. *Guarea cedrata* is known as White guarea, Scented guarea, Bossé, Obobonufua, or Pink mahogany. *Guarea thompsonii* is referred to as Black guarea and Obobonekwi.

**Guarea thompsonii** transverse surface, × 3

**Guarea cedrata** longitudinal surface, × 1

## Anatomy

Figures A–F

Growth rings not present. Wood diffuse-porous. Vessels mainly in radial multiples of 2–4(–5). Simple perforation plates. Alternate and minute intervessel pitting (Fig. D). Vessel-ray pits with distinct borders and similar in size and shape to intervessel pits. Fibre walls thick to very thick. Septate fibres present. Paratracheal axial parenchyma confluent (forming irregular bands, 3–4 cells wide), and aliform. The banded parenchyma is wavy in *G. cedrata* (Fig. A) and straighter and more abundant in *G. thompsonii* (Fig. B). Rays not storied, mainly uniseriate and biseriate (occasionally up to 3 cells wide), and up to 20 cells high (Fig. C). Rays heterocellular with one row (occasionally up to 4 rows) of square and/or upright marginal cells, and homocellular (Fig. E). Prismatic crystals present in chains in chambered axial parenchyma cells (Fig. F). Silica bodies present in ray cells, more commonly in *G. cedrata* than *G. thompsonii*.

The wood anatomical references to *G. cedrata* and *G. thompsonii* include Record (1924), Kribs (1930), Rendle (1938), Lebacq and Istas (1950), Metcalfe and Chalk (1950), Fouarge *et al.* (1953), Normand (1955), Kribs (1959), Brazier and Franklin (1961), Lebacq (1963), Pennington and Styles (1975), Normand and Paquis (1976), and the InsideWood website (2004 onwards).

## Diagnostic characteristics

The wood anatomy of *G. cedrata* and *G. thompsonii* is very similar, however the parenchyma in *G. thompsonii* occurs in straight bands, compared to the wavy and often discontinuous bands present in *G. cedrata*.

*Guarea* is easily separable from true mahogany, due to the aliform and paratracheal bands of parenchyma, compared with the apotracheal bands in *Swietenia*. The banded parenchyma in *Guarea* is similar to that seen in *Dysoxylum* (Fig. A, p. 41). The rays in *Guarea* are distinct from those in *Swietenia* as they are narrow and are both heterocellular and homocellular. Finally, the silica bodies present in *Guarea* are a distinctive characteristic, as they occur in few Meliaceae mahoganies (*Entandrophragma candollei* (p. 44), *Aphanamixis polystachya* (p. 12), and *Chisocheton divergens* (pp. 32–33)).

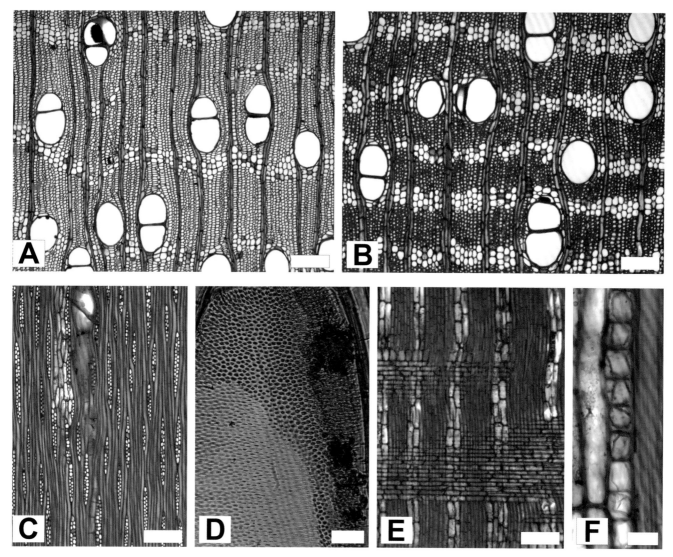

**Figures A, C, D and F** *Guarea cedrata*, **Figures B and E** *Guarea thompsonii*. **A and B.** TS, paratracheal parenchyma confluent, forming irregular marginal bands. Bands wavy in *G. cedrata* (Fig. A) straighter in *G. thompsonii* (Fig. B). **C.** TLS, short uniseriate and biseriate rays, not storied. **D.** TLS, intervessel pits alternate and minute. **E.** RLS, rays heterocellular and homocellular. **F.** RLS, chambered prismatic crystals in axial parenchyma. Figs. A, B, C, E, scale bar = 200 μm, Figs. D and F, scale bar = 20 μm.

# *Khaya* A.Juss.

## Introduction

The genus *Khaya* is very closely related to *Swietenia*, and the timber is very similar in appearance. Historically, trade in *Khaya* from tropical Africa increased when mahogany became a popular wood and as sources of *Swietenia* became more difficult to obtain. *Khaya* has been commercially harvested since the late 1800s, and is one of the major commercial timbers of tropical Africa. *Khaya ivorensis* is the most commercially important species, and is heavily exploited as a result.

## Natural distribution

There are seven species of *Khaya*, five in tropical Africa and two in Madagascar and the Comores islands. The focus here is on the tropical African species, which are more likely to appear in the worldwide timber trade. The species distributions are as follows:

*Khaya anthotheca* extends from Guinea south to Angola and east to Uganda.

*Khaya grandifoliola* occurs from Guinea to Zaire, Uganda and Sudan.

*Khaya ivorensis* is distributed along the coastal rainforests of west Africa from the Ivory Coast south to Angola.

*Khaya nyasica* occurs more inland, in Zaire, Zambia, Uganda and Tanzania.

*Khaya senegalensis* extends across Africa from Senegal to Sudan.

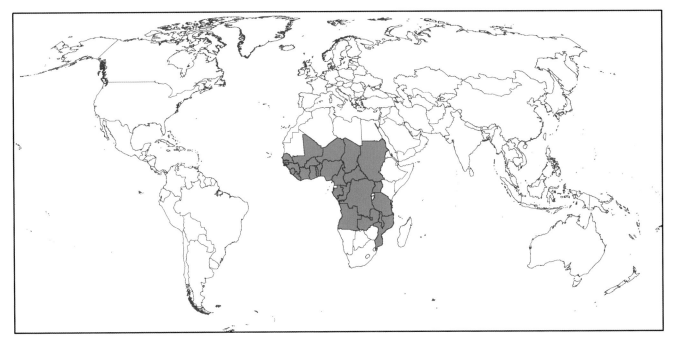

Countries in which *Khaya anthotheca*, *K. grandifoliola*, *K. ivorensis*, *K. nyasica* and *K. senegalensis* occur naturally

# K. anthotheca C.DC., K. grandifoliola C.DC., K. ivorensis A.Chev., K. nyasica Stapf ex Baker f. and K. senegalensis A.Juss.

## Wood

All seven *Khaya* species are indistinguishable based on their wood structure. The wood is very similar to that of *Swietenia*, but with less variation in colour, and is typically red/brown in appearance. By viewing the transverse surface of the timber with the naked eye, there are no bands of axial parenchyma which are often visible in *Swietenia*.

## Uses

The wood properties of *Khaya* timber mean that it is often used as a good substitute for *Swietenia*. Among other applications, *Khaya* timber is used in furniture, panelling, flooring, and veneer production.

## Common names

*Khaya* is generally referred to as African mahogany or Acajou d'Afrique, and common names exist for individual species based upon country of origin, language or local names.

For example, *Khaya anthotheca* is known as White mahogany, Ugandan mahogany or Krala. *Khaya grandifoliola* is known as Heavy African mahogany, or Broad leaved mahogany. *Khaya ivorensis* has many common names including Benin wood, Nigerian mahogany, Red khaya, Grand bassam, Degema mahogany, Lagos mahogany, and Takoradi mahogany. Mozambique mahogany, Nyasaland mahogany and Umbaua are specific to *K. nyasica*. *Khaya senegalensis* is known as Dry zone mahogany, Senegal mahogany and Guinea mahogany.

***Khaya anthotheca*** transverse surface, × 4.5

***Khaya anthotheca*** longitudinal surface, × 1

## Anatomy

Figures A–F

Although some variation occurs in the wood anatomy of each species, the differences are minimal and often occur in different samples of the same species. Therefore the wood anatomy is described at the genus level.

Indistinct growth rings present. Wood diffuse-porous. Vessels mainly in short radial multiples, solitary vessels also present. Simple perforation plates. Minute and alternate intervessel pits (Fig. D). Vessel-ray pits with distinct borders and similar in size and shape to intervessel pits. Fibre walls thin to very thick. Septate fibres present. Paratracheal parenchyma mainly vasicentric (Figs. A and B), and also scanty (Fig. A). Bands of parenchyma very rare. Rays not storied, and of two distinct sizes (Fig. C): short uniseriate and biseriate rays up to 6 cells high, and tall multiseriate rays (up to 6 cells wide and up to 30 cells high) with large upright marginal cells. Rays heterocellular, with one to several marginal rows (4 or more) of upright and/or square cells (Figs. E and F). Sheath cells occasionally present. Prismatic crystals present in upright ray cells (Fig. F).

The wood anatomy of *Khaya* is described in Moeller (1876), Dixon (1919), Koehler (1922), Kribs (1930), Panshin (1933), Rendle (1938), Eggeling and Harris (1939), Lebacq and Istas (1950), Metcalfe and Chalk (1950), Normand (1955), Kribs (1959), Brazier and Franklin (1961), Lebacq (1963), Wagenführ and Steiger (1963), Normand and Paquis (1976), Sosef *et al.* (1998), Negi *et al.* (2003), and the InsideWood website (2004 onwards).

## Diagnostic characteristics

The rays of two distinct sizes with large upright cells at the margins are a key characteristic of *Khaya*, and are rarely seen in other Meliaceae mahoganies. Although large marginal cells may also occur in *Swietenia*, rays of two distinct sizes do not. In addition, the lack of banded apotracheal parenchyma and the abundance of vasicentric paratracheal parenchyma in *Khaya* makes it possible to separate it from *Swietenia*.

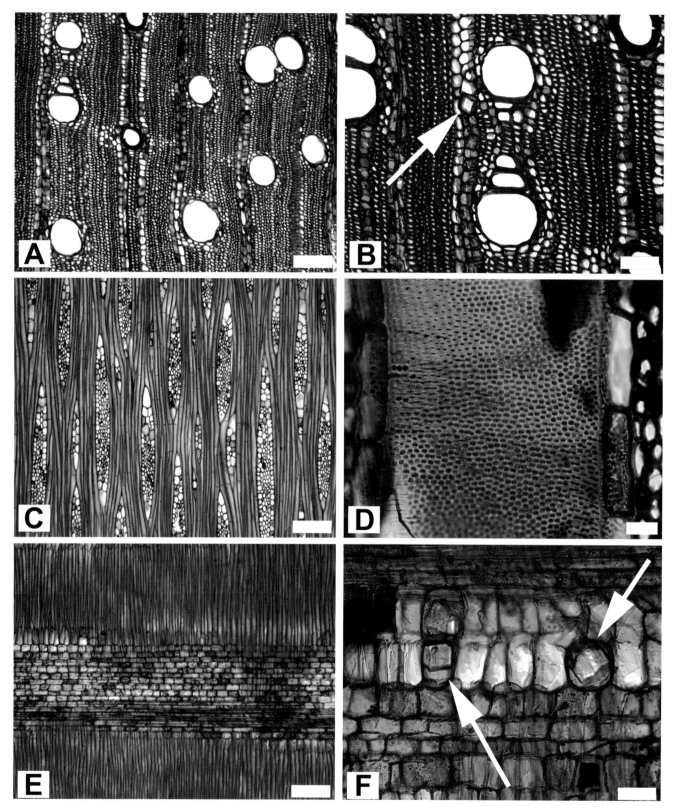

**Figures A–F *Khaya ivorensis*. A.** TS, wood diffuse-porous, showing scanty paratracheal and vasicentric parenchyma. **B.** TS, vasicentric paratracheal parenchyma surrounding a vessel (note also arrow showing prismatic crystal in ray cell). **C.** TLS, short, uniseriate and biseriate rays, and tall multiseriate rays with large upright cells at margins, septate fibres. **D.** TLS, minute, alternate intervessel pitting. **E.** RLS, heterocellular ray, with a single row of upright cells. **F.** RLS, prismatic crystals in upright ray cells (see arrows). Figs. A, C, E, scale bar = 200 μm; Fig. B, scale bar = 100 μm; Fig. D, scale bar = 20 μm; Fig. F, scale bar = 50 μm.

# *Lovoa trichilioides* Harms

### Introduction

Previously known as *Lovoa klaineana, L. trichilioides* is a mahogany timber from Africa. *Lovoa swynnertonii*, the only other species in the genus, is rarely traded on the international market, whereas *L. trichilioides* is frequently traded. In the Congo, *L. trichilioides* is one of the principal timber species (IUCN Red List).

### Natural distribution

*Lovoa trichilioides* is distributed across west Africa and extends from Sierra Leone through the Ivory Coast and Cameroon, and south to the Congo and Zaire.

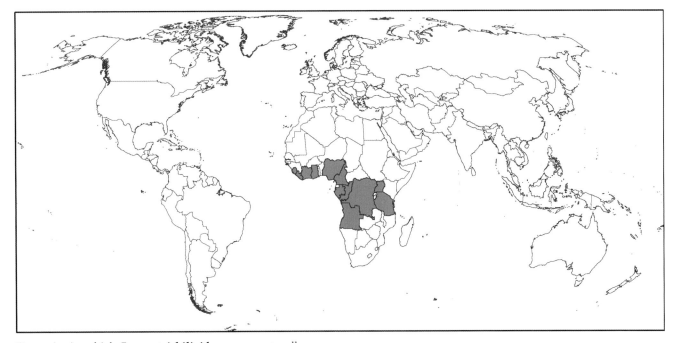

Countries in which ***Lovoa trichilioides*** occurs naturally

## Wood

*Lovoa trichilioides* wood is quite different in colour from most of the other mahoganies, as it is light brown as opposed to darker red/brown (the wood is similar to that of *Turraeanthus africanus* (p. 75)). In contrast to the light colour of the timber, dark lines of marginal traumatic canals are often visible to the naked eye. The abundant traumatic resin canals are characteristic of this genus, and occur far more regularly than they do in other Meliaceae.

## Uses

*Lovoa trichilioides* shares many of the wood properties of other mahoganies, and is mainly used as a decorative timber in furniture and cabinet production as well as a veneer. The light coloured timber may be dyed to resemble true mahogany.

## Common names

*Lovoa trichilioides* is widely known as African walnut, due to its resemblance in colour to walnut (*Juglans* species). In addition the timber is known as Dibetou, Alona wood, Congowood, Tigerwood, Nigerian walnut and Benin walnut.

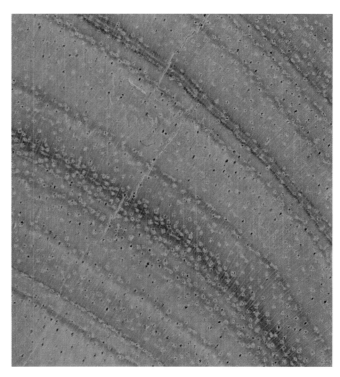

***Lovoa trichilioides*** transverse surface, × 5

***Lovoa trichilioides*** longitudinal surface, × 1

## Anatomy

Figures A–E

Growth rings not present. Wood diffuse-porous. Vessels mainly in radial multiples of 2–4, also solitary. Perforation plates simple. Intervessel pits alternate and minute (Fig. C), to small. Vessel-ray pits with distinct borders and similar in size and shape to intervessel pits. Fibre walls thin to thick. Septate fibres occasionally present. Apotracheal axial parenchyma occasional. Paratracheal axial parenchyma vasicentric (Fig. A), to aliform (winged and lozenge-shaped). Rays occasionally irregularly storied, 2–4 cells wide, up to 20 cells high (Fig. B). Rays homocellular (Fig. D). Prismatic crystals abundant, present in the axial parenchyma in long chains of chambered cells (Fig. E). Tangential lines of traumatic resin canals usually present (Fig. A).

Wood anatomical references to *Lovoa* include Groom (1926), Kribs (1930), Chalk *et al.* (1933), Panshin (1933), Rendle (1938), Lebacq and Istas (1950), Metcalfe and Chalk (1950), Normand (1955), Kribs (1959), Brazier and Franklin (1961), Lebacq (1963), Fouarge and Gérard (1964), Normand and Paquis (1976), and the InsideWood website (2004 onwards).

## Diagnostic characteristics

The wood anatomy of *L. trichilioides* is quite distinct from that of *Swietenia* and the other mahogany timbers, due to the presence of traumatic resin canals. Traumatic canals occur sporadically in the majority of Meliaceae mahoganies, occasionally occurring in *Entandrophragma* (pp. 44–45), *Melia* (p. 60), and *Azadirachta* (pp. 16–17). They frequently occur in *Lovoa*, making the characteristic useful in identification.

In addition, *L. trichiliodes* may be differentiated from other mahogany woods by the combination of homocellular rays, long chains of prismatic crystals in abundance, vasicentric to aliform parenchyma, and minute to small intervessel pits.

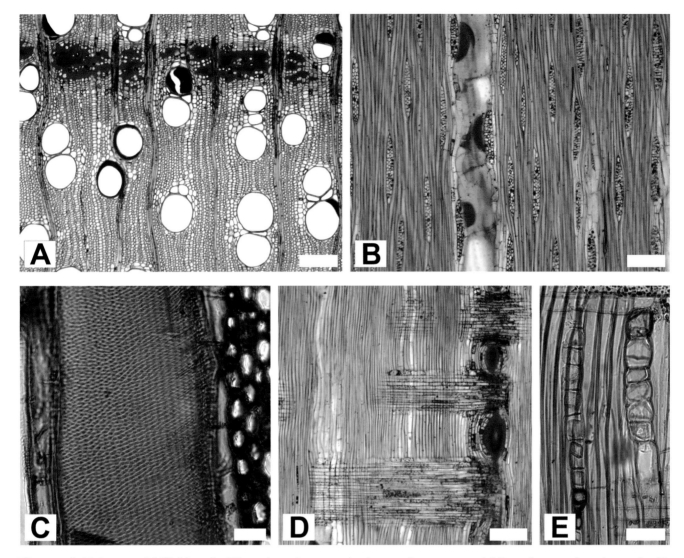

**Figures A–E** *Lovoa trichilioides*. **A.** TS, vasicentric paratracheal parenchyma, tangential line of traumatic resin canals. **B.** TLS, rays up to 4 cells wide and 20 cells high. **C.** TLS, intervessel pits minute and alternate. **D.** RLS, homocellular rays. **E.** RLS, long chains of prismatic crystals in chambered axial parenchyma cells. Figs. A, B, D, scale bar = 200 μm, Fig. C = 20 μm, Fig. E = 50 μm.

# *Melia* L.

## Introduction

*Melia* is one of the major commercial mahogany timbers from Indo-Malaysia, and *M. azedarach* and *M. composita* are commonly traded internationally. *Melia* is very closely related to *Azadirachta*, and the two are often confused due to their similarity and the use of the same common names for both. Meliaceae is named after *Melia*, the type genus of this family.

## Natural distribution

*Melia* exists in plantations from Mexico to Argentina, as well as in South-East Asia and Australia. The natural distribution differs between the two species, extending widely across Asia and South-East Asia.

***Melia azedarach*** is found in India, China, Vietnam, Thailand and Australia. It has been introduced in Africa (Uganda, Tanzania) and has become naturalized in the Americas.

***Melia composita*** is restricted to India, Sri Lanka and Bhutan.

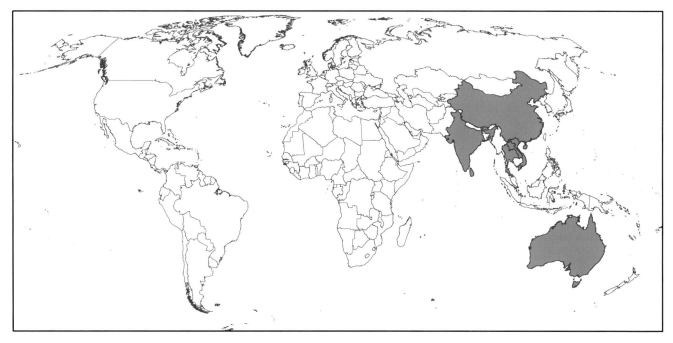

Countries in which ***Melia azedarach*** and ***Melia composita*** occur naturally

# *M. azedarach* L. and *M. composita* Willd.

## Wood

The wood of *Melia* exhibits the same golden brown to pink/red brown colour seen in many other mahogany timbers, and upon ageing becomes lighter. The semi-ring and ring porous nature of the vessels creates darker-brown streaks in the wood.

## Uses

The wood is used as a quality veneer and to make musical instruments, cabinets, cigar boxes, and small items such as toys and boxes. The pulp is used to make paper.

## Common names

*Melia azedarach* is known as Persian lilac, Chinaberry, Carolina mahogany, Pride of India and Paternostertree. *Melia composita* is known as Malabar nim wood or Lunumidella.

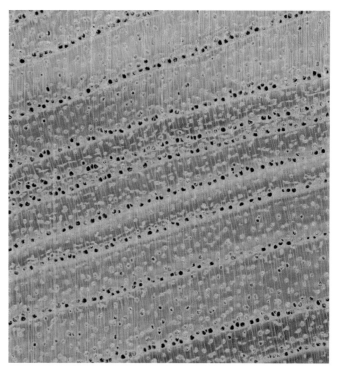

***Melia composita*** transverse surface, × 4

***Melia composita*** longitudinal surface, × 1

## Anatomy

Figures A–G

Growth rings present (Figs. A and B). Wood diffuse to semi-ring-porous in *M. composita* (Fig. A), and ring-porous in *M. azedarach* (Fig. B). Vessels in a radial or diagonal pattern in latewood of *M. azedarach* (Fig. B). Vessels solitary, occasionally in radial multiples of two in *M. composita* (Fig. A), and abundant vessel clusters in *M. azedarach* (Fig. B). Simple perforation plates. Intervessel pits small to medium and alternate (Fig. D). Vessel-ray pits with distinct borders and similar in size and shape to intervessel pits. Striations in vessel walls resembling helical thickenings sometimes visible (Fig. E). Fibre walls thin to thick. All fibres non-septate. Apotracheal axial parenchyma present in indistinct marginal bands (Fig. B), mainly up to 4 cells wide. Paratracheal axial parenchyma scanty to vasicentric, becoming confluent in *M. azedarach* (Fig. B). Rays not storied, 1–6 cells wide, and ranging in height from 6 to 30 cells (occasionally up to 40) (Fig. C). Rays heterocellular with one row of square and/or upright marginal cells (Fig. F) and homocellular. Prismatic crystals abundant in *M. azedarach*, in chains of chambered axial parenchyma cells (Fig. G). Crystals occasionally present in *M. composita* in ray cells and axial parenchyma. Tangential lines of traumatic resin canals occasional.

The wood anatomy of *Melia* is described in Moeller (1876), Kribs (1930), Pearson and Brown (1932), Metcalfe and Chalk (1950), Brazier and Franklin (1961), Ghosh *et al.* (1963), Pennington and Styles (1975), Normand and Paquis (1976), Datta and Samanta (1983), Nair (1987), Nair (1991), Negi *et al.* (2003), and the InsideWood website (2004 onwards).

## Diagnostic characteristics

The differences between these two species are very significant. The diagonal pattern of vessels in *M. azedarach* is not found in *M. composita* or any other mahoganies. In addition, the great abundance of prismatic crystals in *M. azedarach* is a distinctive characteristic, especially in comparison with *M. composita*.

*Melia composita* is easily distinguishable from *Swietenia* and most other mahoganies due to the combination of semi-ring to ring-porous wood (also seen in *Cedrela* (pp. 28–29), and *Toona* (pp. 72–73)), small intervessel pits, non-septate fibres, unstoried rays, the combination of heterocellular and homocellular rays, and striations in vessel walls. *Melia azedarach* is unique in having the diagonal vessel arrangement and is unlikely to be confused with any other mahogany.

It is difficult to distinguish between the anatomy of *M. composita* and *Azadirachta* (pp. 16–17). However, the distinctive vessel pattern in *M. azedarach* is not seen in *Azadirachta*, and banded parenchyma is far more abundant in *Azadirachta*.

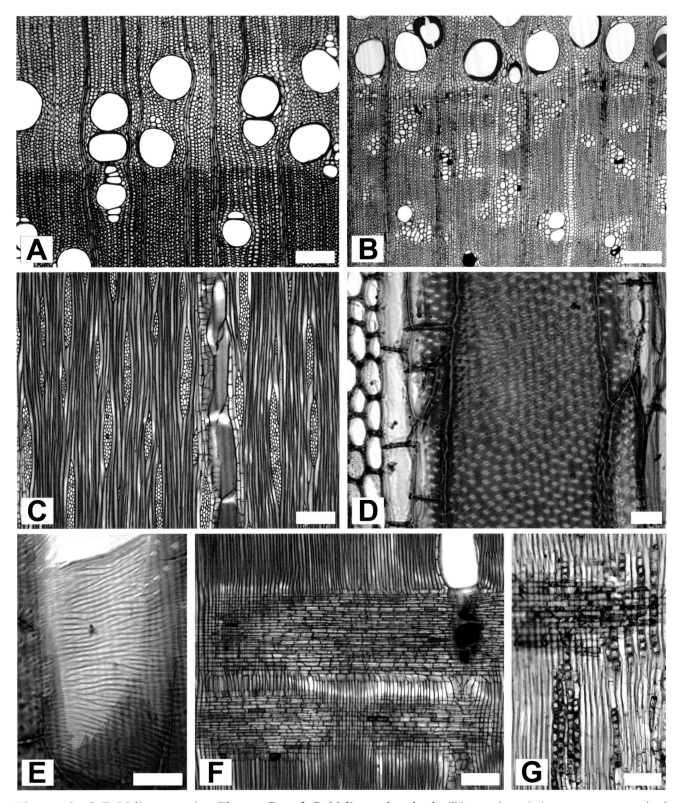

**Figures A, C–F** *Melia composita*, **Figures B and G** *Melia azedarach*. **A.** TS, wood semi-ring-porous, paratracheal parenchyma scanty. **B.** TS, wood ring-porous, latewood vessels in an indistinct diagonal pattern, paratracheal parenchyma scanty, vasicentric and confluent. **C.** TLS, mainly tall multiseriate rays, not storied. **D.** TLS, alternate intervessel pitting small to medium. **E.** TLS, striations in the vessel wall resembling helical thickenings. **F.** RLS, rays heterocellular. **G.** RLS, abundant prismatic crystals in the axial parenchyma, in chains of chambered cells. Figs. A–C and F, scale bar = 200 μm, Figs. D and E, scale bar = 20 μm, Fig. G, scale bar = 100 μm.

# *Pseudocedrela kotschyi* Harms

## Introduction

*Pseudocedrela kotschyi* is the only species of the genus, though historically various other species have been described and disregarded. The timber is not highly exported and is much less likely to be found on the commercial market than other mahogany woods, even though it closely resembles the wood colouring and anatomy of *Swietenia*.

## Natural distribution

*Pseudocedrela kotschyi* has a wide distribution in tropical Africa, and ranges across the continent from Nigeria and the Cameroon, into Chad, Sudan and Uganda.

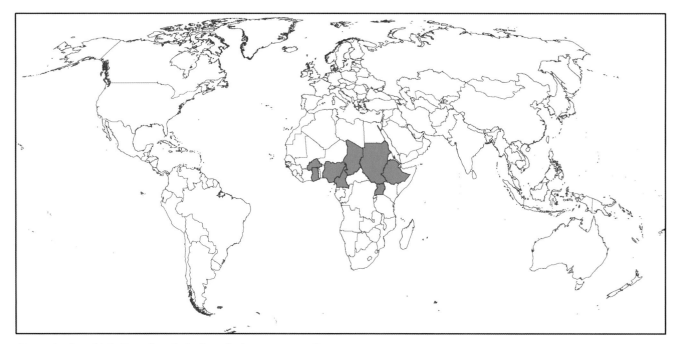

Countries in which ***Pseudocedrela kotschyi*** occurs naturally

## Wood

The deep red/brown colour of *P. kotschyi* greatly resembles that of *Swietenia* mahogany.

## Uses

*Pseudocedrela kotschyi* is used locally as timber for construction and canoes, but is unlikely to be found on the international market. The roots, leaves and bark are used for medicinal purposes.

## Common names

There are few common names associated with *P. kotschyi* as it is of limited commercial importance, but it is known as Dry zone cedar or Hard cedar-mahogany.

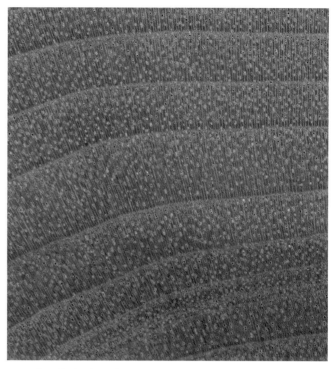

***Pseudocedrela kotschyi*** transverse surface, × 4

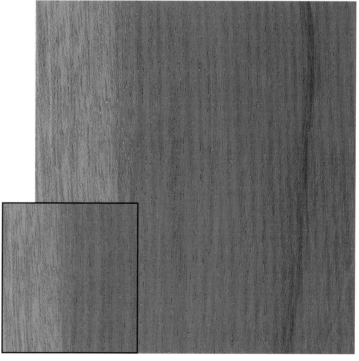

***Pseudocedrela kotschyi*** longitudinal surface, × 1 (inset) and × 1.25

## Anatomy

Figures A–E

Growth rings present. Wood diffuse-porous. Vessels mainly in radial multiples of 2–3, also solitary (Fig. A). Simple perforation plates. Alternate and minute intervessel pits (Fig. C). Vessel–ray pits with distinct borders and similar in size and shape to intervessel pits. Fibre walls thin to thick. Septate fibres present occasionally. Apotracheal axial parenchyma marginal and banded, between 3 and 4 cells wide (Fig. A), and occasionally diffuse. Paratracheal axial parenchyma scanty and vasicentric (Fig. A). Rays storied, 2–4 cells wide, up to 20 cells high, and with tall marginal cells (Fig. B). Rays heterocellular with one row of square and/or upright marginal cells (Fig. D). Prismatic crystals present in upright ray and axial parenchyma cells (Fig. E).

References to the wood anatomy of *Pseudocedrela* include Kribs (1930), Chalk *et al.* (1933), Panshin (1933), Metcalfe and Chalk (1950), Brazier and Franklin (1961), and the InsideWood website (2004 onwards).

## Diagnostic characteristics

The wood anatomy of this species is very similar to that of *Swietenia*, primarily due to the banded axial parenchyma that is found in both mahoganies. The paratracheal parenchyma is similar in both (scanty and vasicentric, becoming more vasicentric in *P. kotschyi*), and storied rays of similar sizes are a common feature. However, the diffuse parenchyma and the tall marginal ray cells that occur in *P. kotschyi* are not a common feature of *Swietenia*, and such ray cells are more likely to occur in *Khaya* (pp. 52–53). *Pseudocedrela kotschyi* has banded parenchyma, and *Khaya* does not.

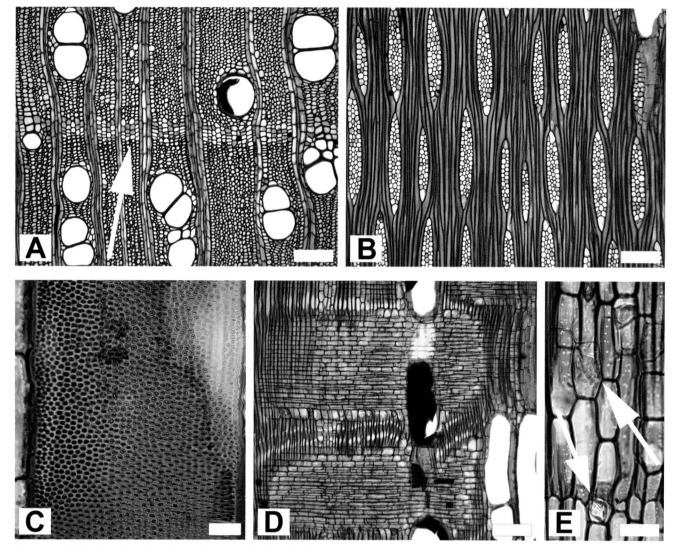

**Figures A–E** *Pseudocedrela kotschyi.* **A.** TS, band of marginal apotracheal parenchyma (see arrow), vasicentric and scanty paratracheal parenchyma. **B.** TLS, rays storied, with tall marginal cells. **C.** TLS, intervessel pits alternate and minute. **D.** RLS, rays heterocellular with large procumbent or upright cells at ray margins. **E.** RLS, non-chambered prismatic crystals in the axial parenchyma (see arrows). Figs. A, B, D, scale bar = 200 μm, Fig. C = 20 μm, Fig. E = 50 μm.

# *Swietenia* Jacq.

## Introduction

The three species of *Swietenia* are the original sources of true mahogany timber, and were the first to be traded under the name mahogany. Following the success of *Swietenia* as a high-quality timber, other often unrelated woods have also been called 'mahogany'.

*Swietenia macrophylla* is one of the world's most valuable tropical timbers. *Swietenia mahagoni* and *S. humilis* are now considered to be commercially extinct, having been threatened with extinction by over-harvesting and illegal logging. Legal protection of these two species by placing them on CITES Appendix II was imposed in 1975 (*S. humilis*) and 1992 (*S. mahagoni*). *Swietenia macrophylla* was moved from Appendix III to Appendix II in November 2003 to afford it the same level of protection.

## Natural distribution

Although there are mahogany plantations (primarily of *S. macrophylla*) in South-East Asia, the natural distribution is confined to tropical America.

**Swietenia humilis** is distributed in dry areas along the Pacific coast of Mexico and Central America, in tropical deciduous forest.

**Swietenia macrophylla** extends from the lower end of Mexico, through Central America and into South America (Colombia, Venezuela, Peru, Bolivia, and Brazil) in seasonally wet areas of high rainfall.

**Swietenia mahagoni** occurs in the Caribbean and the tip of southern Florida in a strongly seasonal climate.

Countries in which **Swietenia humilis, S. macrophylla** and **S. mahagoni** occur naturally

# *S. humilis* Zucc., *S. macrophylla* King and *S. mahagoni* (L.) Jacq.

## Wood

The colour of the wood can vary from light yellow to brown, and from brown to red, depending upon the incidence of resin and gum deposits. Large vessels can be seen on the transverse surface with the naked eye, and distinct bands of axial parenchyma can often also be seen.

## Uses

*Swietenia macrophylla* is now the only commercially available timber of the genus, following the depletion of *S. mahagoni* and *S. humilis*. It is a high quality timber used in the production of furniture, musical instruments, coffins, panelling and as a veneer.

## Common names

*Swietenia* is commonly known as Mahogany, American mahogany, Caoba in Spanish, and Acajou in French. There are many other common names in use that refer to each species of *Swietenia*, usually based upon the country of origin, local language, or trade.

For example, *S. humilis* is known as Dry zone mahogany, Pacific coast mahogany, Honduras mahogany or Mexican mahogany, and *S. mahagoni* as Caribbean mahogany, West Indian mahogany, Spanish mahogany, and Small leaved mahogany. *Swietenia macrophylla* has many common names due to its volume in trade and wide distribution, but is commonly referred to as Bigleaf mahogany, Honduras mahogany, Brazilian mahogany, and Mara or Mogno in Spanish.

***Swietenia macrophylla*** transverse surface, × 2

***Swietenia mahagoni*** longitudinal surface, × 1

## Anatomy

Figures A–G

The anatomical description given here is to the genus level. Differences between the three species are inconsistent, and it is impractical to identify a *Swietenia* wood sample to species.

Growth rings present but indistinct. Wood diffuse-porous. Vessels solitary and in short radial multiples of 2–4, occasionally up to 6. Perforation plates simple. Intervessel pits minute and alternate (Fig. D). Vessel-ray pits with distinct borders and similar in size and shape to intervessel pits. Fibre walls thin to thick. Septate fibres in varying abundance (Fig. C), in combination with non-septate fibres. Apotracheal axial parenchyma banded and marginal, up to 5 cells wide (Fig. A), sometimes indistinctly (Fig. B). Paratracheal axial parenchyma scanty (Fig. A), and very occasionally vasicentric. Rays regularly storied in *S. macrophylla* (Fig. C), but often irregularly storied in the other two species. Rays 2–5 cells wide, 12–18 cells high. Rays heterocellular with 1–2 rows of upright and/or square cells (Fig. E), with prismatic crystals in upright ray cells (Fig. F), and occasionally in the axial parenchyma (Fig. G).

The wood anatomy of *Swietenia* is described in Koehler (1922), Record (1924), Kribs (1930), Panshin (1933), Rendle (1938), Metcalfe and Chalk (1950), Kribs (1959), Ghosh *et al.* (1963), Datta and Samanta (1983), Soerianegara and Lemmens (1993), and Negi *et al.* (2003).

References to the wood anatomy of either or both *S. macrophylla* and *S. mahagoni* include Moeller (1876), Dixon (1919), Brazier and Franklin (1961), Wagenführ and Steiger (1963), Gaiotti de Peralta and Edlmann Abbate (1981), Détienne and Jacquet (1983), Donaldson (1984), Nair (1987), Nair (1991), Acevedo Mallque and Kikata (1994), Dünisch *et al.* (2002) and the InsideWood website (2004 onwards). Wood anatomical descriptions of *S. humilis* are rare, however one is provided in Barajas-Morales and Gomez (1989).

## Diagnostic characteristics

The two anatomical features that reliably separate *Swietenia* from the other 'mahoganies' are the banded apotracheal parenchyma and storied multiseriate rays (regularly and irregularly). The African genus *Khaya* (pp. 52–53) is most similar to *Swietenia*, but has rays of two distinct sizes and no banded apotracheal parenchyma.

Few other Meliaceae mahoganies have the apotracheal bands of parenchyma that are characteristic of *Swietenia*, but they do occur in *Azadirachta* (pp. 16–17), *Chukrasia tabularis* (pp. 36–37), and *Pseudocedrela kotschyi* (pp. 64–65). When the other wood anatomical characteristics are taken into account for these taxa, they are clearly different from the true mahogany, *Swietenia*.

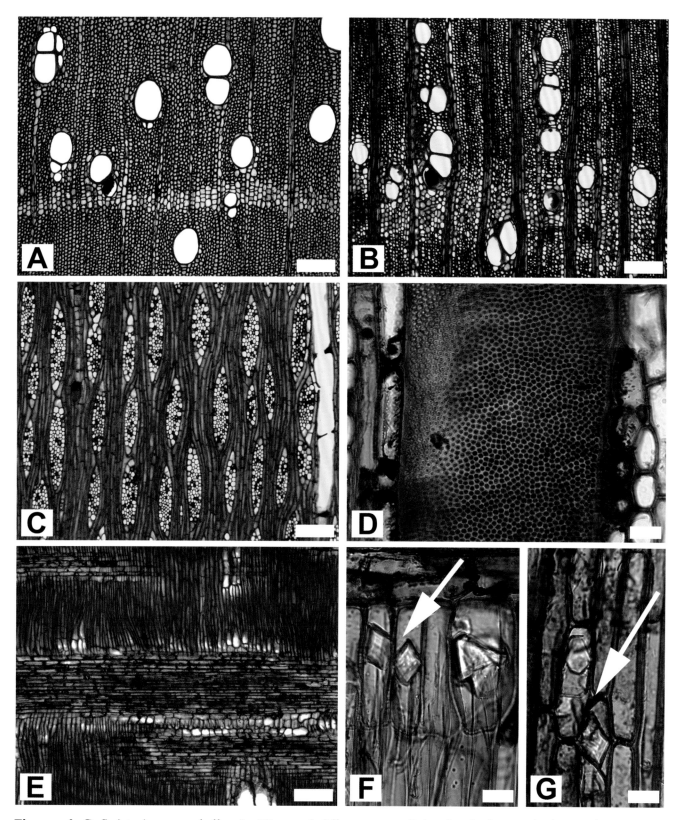

**Figures A–G** *Swietenia macrophylla.* **A.** TS, wood diffuse porous, distinct band of apotracheal parenchyma, scanty paratracheal parenchyma. **B.** TS, indistinct band of apotracheal parenchyma. **C.** TLS, rays storied. Septate fibres. **D.** TLS, minute and alternate intervessel pitting. **E.** RLS, heterocellular ray, with one to two rows of upright/square marginal cells. **F.** RLS, prismatic crystals in upright ray cells (see arrow). **G.** RLS, prismatic crystals in axial parenchyma (see arrow). Figs. A–C and E, scale bar = 200 μm, Figs. D, F and G, scale bar = 20 μm.

# *Toona ciliata* M.Roem.

## Introduction

There are approximately six species of *Toona* from India, South-East Asia and Australia. Historically, *Toona* and *Cedrela* (pp. 26–29) were often confused, as the two genera are very closely related and almost identical morphologically and anatomically. Synonyms of *T. ciliata* include *T. australis*, *T. microcarpa*, *T. ternatensis*, and confusingly, *Cedrela toona*. The American species are all now considered as *Cedrela*, and the South-East Asian, Indian and Australian species as *Toona*.

*Toona* is a valuable timber that has been heavily harvested and traded, and *T. ciliata* is one of the major traded species.

## Natural distribution

*Toona ciliata* is widely distributed across India, China and Australia. Plantations exist in tropical America.

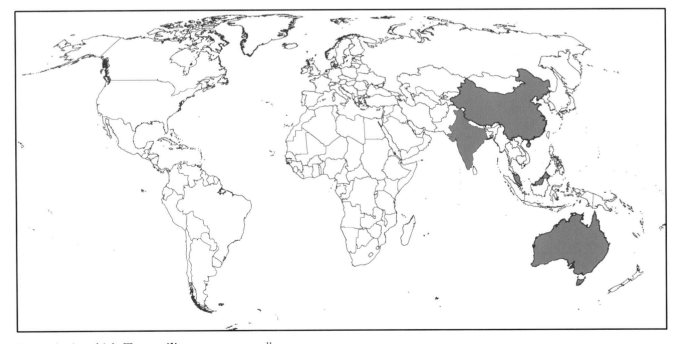

Countries in which ***Toona ciliata*** occurs naturally

## Wood

The wood is pale to dark red/brown, often 'brick red' in colour, with dark streaks as a result of the semi-ring porous growth rings.

## Uses

The timber is used in the construction of houses and ships, and for lighter construction of furniture and musical instruments. The bark is used for its medicinal properties, and a dye can be extracted from the flowers. *Toona ciliata* is often planted as an ornamental or street tree.

## Common names

*Toona* in general is known as Toon, and *T. ciliata* in particular is known as Indian mahogany, Red cedar (this name is also used for *Cedrela* (p. 27)), Burmese cedar, Yomham, Suren kapar, Malapoga, Kukoru and Thitkado.

***Toona ciliata*** transverse surface, × 2

***Toona ciliata*** longitudinal surface, × 1

## Anatomy

Figures A–G

Growth rings distinct, wood diffuse to semi-ring-porous (Fig. A), and occasionally ring-porous. Vessels mainly solitary, occasionally in radial multiples of up to three. Perforation plates simple. Intervessel pits minute to small and alternate (Fig. D). Vessel-ray pits with distinct borders and similar in size and shape to intervessel pits. Fibre walls thin to thick. Fibres non-septate. Apotracheal parenchyma diffuse and diffuse-in-aggregates, and in initial bands in association with the growth ring (Fig. A). Paratracheal parenchyma scanty and vasicentric (Fig. A). Rays heterocellular with one row of square and/or upright marginal cells (Fig. E), and not storied. Rays 1–3 cells wide (Fig. B), and up to 10 cells wide (Fig. C). Rays up to 20 cells high. Prismatic crystals present in square and upright ray cells and chambered axial parenchyma (Fig. F). Druses very occasional in non-chambered axial parenchyma (Fig. G). Tangential lines of traumatic resin canals occasionally present.

References for *Toona ciliata* include Pearson and Brown (1932) (as *Cedrela toona*), Ghosh *et al.* (1963), Datta and Samanta (1983) (*Toona*), Espinoza de Pernia (1987), Lemmens *et al.* (1995), Negi *et al.* (2003), and the InsideWood website (2004 onwards).

## Diagnostic characteristics:

The semi-ring porous wood of *T. ciliata* differentiates it from *Swietenia*, as does the diffuse apotracheal parenchyma. When present, druses aid in the separation of *T. ciliata* from *Swietenia* and many other mahogany woods.

The wood anatomy of *Toona* greatly resembles that of *Cedrela* (pp. 28–29), and unless the geographical origin is known differentiating between the two is probably impractical.

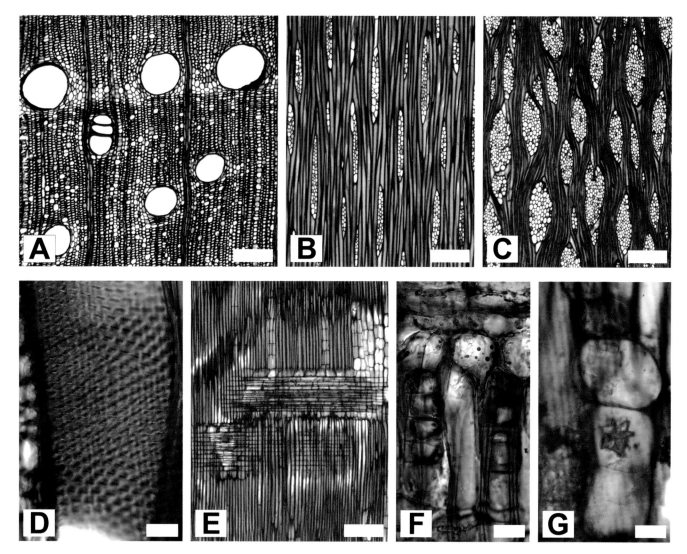

**Figures A–G** *Toona ciliata*. **A.** TS, wood semi-ring porous, apotracheal parenchyma diffuse and diffuse-in-aggregates, and forming a marginal (initial) band. Paratracheal parenchyma scanty and vasicentric. **B and C.** TLS, rays 1–3 cells wide (Fig. B) and up to 10 cells wide (Fig. C), and not storied. **D.** TLS, intervessel pits small and alternate. **E.** RLS, heterocellular rays with one row of square and/or upright marginal cells. **F.** RLS, prismatic crystals present in chambered axial parenchyma. **G.** RLS, druse in axial parenchyma. Figs. A, B, C, E, scale bar = 200 μm, Figs. D, F, G, scale bar = 20 μm.

# *Turraeanthus africanus* (Welw.) Pellegr.

## Introduction

*Turraeanthus africanus* is an African mahogany, and represents one of two or three species in the genus. The other species are unlikely to be found on the international timber market.

## Natural distribution

*Turraeanthus* is found in tropical west Africa, and is not as widely distributed as some of the other African mahoganies. *Turraeanthus africanus* occurs in Sierra Leone and the Ivory Coast, east to Cameroon and south to Angola.

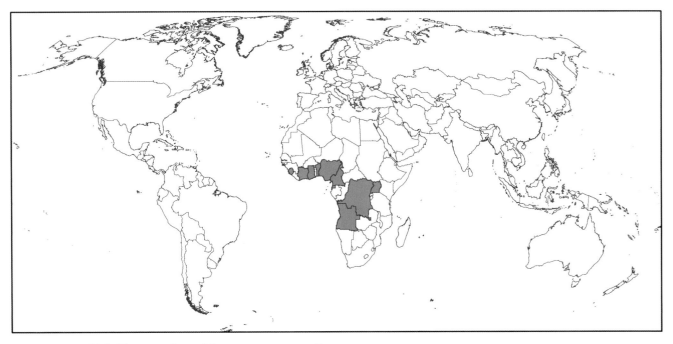

Countries in which ***Turraeanthus africanus*** occurs naturally

## Wood

The colour of *T. africanus* wood is similar to that of *Lovoa* (p. 55), another African mahogany. It is pale yellow/brown in colour and may also have a golden colouring. The sapwood is the same colour. The wood properties are similar to true mahogany and the timber is used for similar purposes.

## Uses

The timber may be used in furniture production and as a veneer, as well as for flooring and musical instruments.

## Common names

*Turraeanthus africanus* is commonly known by the French name Avodire. It is also known as African satinwood, African goldbirch and Lusamba.

***Turraeanthus africanus*** transverse surface, × 5

***Turraeanthus africanus*** longitudinal surface, × 1

**Anatomy**

Figures A–D

Growth rings not seen. Wood diffuse-porous. Vessels mainly solitary, occasionally in radial multiples or clusters of 2–3 (Fig. A). Perforation plates simple. Intervessel pits alternate and minute (Fig. C). Vessel-ray pits with distinct borders and similar in size and shape to intervessel pits. Fibre walls thin to thick. Septate fibres occasionally present. No apotracheal axial parenchyma. Paratracheal axial parenchyma vasicentric and scanty (Fig. A). Rays irregularly storied (Fig. B), 1–3 cells wide, and up to 15 cells high. Rays heterocellular with one row of square and/or upright marginal cells, and homocellular (Fig. D). Prismatic crystals in the axial parenchyma, in chambered cells, in chains (Fig. D).

References to the wood anatomy of *Turraeanthus* include Kribs (1930), Lebacq and Istas (1950), Metcalfe and Chalk (1950), Fouarge *et al.* (1953), Normand (1955), Kribs (1959), Brazier and Franklin (1961), Lebacq (1963) (*T. africanus*), Pennington and Styles (1975), Normand and Paquis (1976), and the InsideWood website (2004 onwards).

## Diagnostic characteristics

The vasicentric parenchyma and absence of banded apotracheal parenchyma distinguish *T. africanus* from *Swietenia*. Anatomically, *T. africanus* more closely resembles *Khaya* (pp. 52–53), but *T. africanus* has smaller rays and long chains of chambered prismatic crystals.

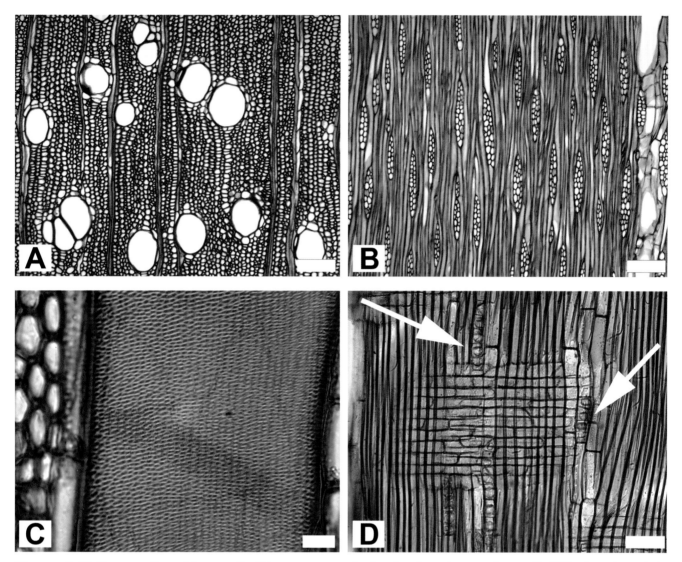

**Figures A–D** *Turraeanthus africanus*. **A.** TS, scanty paratracheal and vasicentric parenchyma. **B.** TLS, short, irregularly storied rays. **C.** TLS, minute, alternate intervessel pits. **D.** RLS, ray homocellular, with chains of prismatic crystals in chambered axial parenchyma cells (see arrows). Figs. A and B, scale bar = 200 μm, Fig. C = 20 μm, Fig. D = 100 μm.

# Wood anatomy of Meliaceae mahoganies

Definitions of the abbreviations used in Table 1. Anatomical characteristics are based upon those detailed in the *IAWA list of microscopic features for hardwood identification* (Wheeler, Baas & Gasson, 1989). Various characteristics have been omitted as they are identical for all taxa studied. All perforation plates are simple, intervessel pits are alternate, and vessel-ray pits have distinct borders and are similar to intervessel pits in size and shape.

## Key: Definitions of wood anatomical characteristics

| Characteristic | Definition | |
| --- | --- | --- |
| **Growth rings** | | |
| Y/N/I | Yes/No/Indistinct | |
| **Porosity** | | |
| R | Ring porous | |
| SR | Semi ring porous | |
| D | Diffuse porous | |
| **Size of intervessel pits** | | |
| Min | Minute | $\leq 4\ \mu m$ |
| S | Small | 4-7 $\mu m$ |
| M | Medium | 7-10 $\mu m$ |
| L | Large | $\geq 10\ \mu m$ |
| **Fibre wall thickness** | | |
| 0 | Thin | |
| 1 | Thick | |
| 2 | Very thick | |
| **Paratracheal axial parenchyma** | | |
| Sc | Scanty | |
| V | Vasicentric | |
| Al | Aliform | |
| R | Reticulate | |
| **Banded axial parenchyma** | | |
| Y/N/I | Yes/No/Indistinct | |
| **Rays storied** | | |
| Y/N/I | Yes/No/Irregular | |
| **Location of crystals/silica bodies** | | |
| RC | Ray cells | |
| Ax | Axial parenchyma | |
| Ch | Chambered | |
| (CS) | Chains | |
| F | Fibres | |

# Table 1: Wood anatomy of Meliaceae mahoganies

| Species | Growth rings Y/N/I | Porosity R/SR/D | Radial multiples (number of vessels) | Size of intervessel pits (alternate) Min/S/M/L | Fibre wall thickness 0/1/2 | Septate fibres Y/N | Apotracheal axial parenchyma Y/N | Paratracheal axial parenchyma Sc/V/Al/R | Axial parenchyma confluent Y/N | Banded axial parenchyma Y/N/I (width in cells) | Rays storied Y/N/I | Ray width in cells | Ray height in cells | Rays Homocellular/Heterocellular | Number rows upright/square cells in heterocellular rays | Sheath cells Y/N | Prismatic crystals Y/N | Location of crystals RC/Ax/Ch/(CS) | Silica bodies Y/N | Location of silica bodies RC/Ax/F | Tangential traumatic resin canals Y/N |
|---|---|---|---|---|---|---|---|---|---|---|---|---|---|---|---|---|---|---|---|---|---|
| *Aglaia argentea* | I | D | 2–5 | Min | 0/1 | Y | N | Al, (Sc) | Y | – | N | 1–2 | 10–30 | He | 1 | N | Y | Ax (CS), Ch | N | – | N |
| *Aphanamixis polystachya* | I | D | 2–5 | S | 0/1 | Y | N | Sc, Al | Y | I(3–5) | N | 1–2 | 3–20 | He | 1–2 | N | Y | Ax (CS), Ch | Y | RC, Ax | N |
| *Azadirachta (A. excelsa, A. indica)* | Y | D | 2–6 | Min–S | 0/1 | N | Y | Sc, V | N | Y (3–6) | N | 2–4 | 10–40 | He/Ho | 1 | N | Y | Ax (CS), Ch | N | – | Y |
| *Cabralea canjerana* | I | D | 2–4 | S | 0/1 | Y | N | Al | Y | Y (3–8) | N | 1–2 | 5–25 | He | 1 | N | N | – | N | – | N |
| *Carapa (C. guianensis, C. procera)* | I | D | 2–3 | Min–S | 0/1 | Y | Y | Sc, V, Al | N | Y (1–3) | I | 1–3 | 6–30 | He | 1–4 | N | Y | RC, Ax, Ch | N | – | N |
| *Cedrela (C. fissilis, C. odorata)* | Y | SR/R | 0 | S–M | 0/1 | Y | Y | Sc, V, Al | N | Y (1–8) | N | 2–3 | 5–30 | He | 1–4 | N | Y | RC, Ax | N | – | N |
| *Chisocheton divergens* | Y | D | 2–4 | Min–S | 0/1 | Y | N | R | N | Y (2–3) | N | 1–3 | 5–20 | He | 1–4 | N | Y | Ax (CS), Ch | Y | RC, Ax, F | N |
| *Chukrasia tabularis* | I | D | 2–3 | Min–S | 0/1 | N | Y | Sc, (V) | N | Y (1–3) | N | 1–3 | 5–15 | He | 1 | N | Y | Ax (CS), Ch | N | – | N |
| *Dysoxylum fraserianum* | N | D | 2–3 | S–M | 1/2 | Y | N | V | Y | Y (3–4) | N | 1–3 | 3–20 | He | 1 | N | Y | Ax (CS), Ch | N | – | N |
| *Entandrophragma (E. angolense, E. candollei, E. cylindricum, E. utile)* | I | D | 2 | Min | 0/1/2 | Y | N | Sc, V | N | Y (1–5) | I,Y | 3–5 | 5–20 | He | 1 | N | Y | RC, Ax (CS), Ch | Y | RC, Ax | Y |
| *Guarea (G. cedrata, G. thompsonii)* | N | D | 2–5 | Min | 1/2 | Y | N | Al | Y | Y (3–4) | N | 1–3 | 5–20 | He/Ho | 1–4 | N | Y | Ax, Ch | Y | RC | N |
| *Khaya (K. anthotheca, K. grandifoliola, K. ivorensis, K. nyasica, K. senegalensis)* | I | D | 2 | Min | 0/1/2 | Y | N | Sc, V | N | N | N | 1–6 | 3–30 | He | 1–4 | Y | Y | RC | N | – | N |
| *Lovoa trichilioides* | N | D | 2–4 | Min–S | 0/1 | Y | Y | V, Al | N | N | I | 2–4 | 5–20 | Ho | – | N | Y | Ax (CS), Ch | N | – | Y |
| *Melia (M. azedarach, M. composita)* | Y | D/SR/R | 2 | S–M | 0/1 | N | Y | Sc, V | Y | Y (1–4) | N | 1–6 | 6–40 | He/Ho | 1–2 | N | Y | RC, Ax (CS), Ch | N | – | Y |
| *Pseudocedrela kotschyi* | I | D | 2–3 | Min | 0/1 | Y | Y | Sc, V | N | Y (3–4) | Y | 2–4 | 20 | He | 1 | N | Y | RC, Ax | N | – | N |
| *Swietenia (S. humilis, S. macrophylla, S. mahagoni)* | I | D | 2–4 (–6) | Min | 0/1 | Y | Y | Sc, V | N | Y (2–5) | I,Y | 2–5 | 12–18 | He | 1–2 | N | Y | RC, (Ax) | N | – | N |
| *Toona ciliata* | Y | D/SR | 2–3 | Min–S | 0/1 | N | Y | Sc, V | N | Y (2–6) | N | 1–10 | 6–20 | He | 1 | N | Y | RC, Ax, Ch | N | – | Y |
| *Turraeanthus africanus* | N | D | 2–3 | Min | 0/1 | Y | N | Sc, V | N | N | I | 1–3 | 5–15 | He/Ho | 1 | N | Y | Ax (CS), Ch | N | – | N |

# NON-MELIACEAE MAHOGANIES

## Table 2: Wood anatomy of non-Meliaceae mahoganies

This table summarises the most useful features for separating these timbers from the Meliaceae mahoganies, and does not include characteristics commonly found in Meliaceae mahoganies. It is not intended as a detailed description of each wood. These taxa have been included as they are known as a type of mahogany, and may be known as such locally, by common or trade name. The anatomical characteristics are based upon those detailed in the *IAWA list of microscopic features for hardwood identification* (Wheeler, Baas & Gasson 1989), and fuller descriptions of many can be found on the InsideWood website (2004 onwards).

| Species | Family | Common name(s) | Geographical distribution | Distinctive wood anatomy |
|---|---|---|---|---|
| *Kiggelaria africana* L. | Achariaceae | Natal mahogany | Africa | – Vessels in a radial pattern (Fig. 1)<br>– Long radial multiples (Fig. 1)<br>– Tyloses (Figs. 1 and 2)<br>– Paratracheal parenchyma rare to absent<br>– Rays of two distinct sizes (Fig. 2)<br>– Vessel-ray pits with much reduced borders to apparently simple: pits rounded or angular and horizontal (scalariform, gash-like) to vertical (palisade) (Fig. 3)<br>– Prismatic crystals in ray cells (Fig. 3, see arrow) |
| *Betula lenta* L. | Betulaceae | Mahogany birch<br>Sweet birch | North America | – Growth rings<br>– Apotracheal parenchyma diffuse (Fig. 4)<br>– No paratracheal parenchyma<br>– Rays of two sizes: large multiseriate and small uniseriate (Fig. 5)<br>– Scalariform perforation plates<br>– No prismatic crystals |
| *Tabebuia donnell-smithii* Rose | Bignoniaceae | White mahogany<br>Primavera | South America, Mexico | – Growth rings (Fig. 6)<br>– Tyloses<br>– Parenchyma vasicentric, lozenge-aliform, confluent banded (Fig. 6)<br>– Rays and axial elements irregularly storied<br>– No prismatic crystals |
| *Aucoumea klaineana* Pierre | Burseraceae | Gaboon mahogany<br>Okoumé | Africa | – Parenchyma rare (scanty) or absent (Fig. 7)<br>– Intervessel pits alternate and large (Fig. 8)<br>– Vessel-ray pits with much reduced borders to apparently simple: pits rounded or angular and horizontal (scalariform, gash-like) to vertical (palisade) (Fig. 9)<br>– No prismatic crystals<br>– Silica bodies in ray cells (Fig. 9, see arrow) |
| *Canarium euphyllum* Kurz | Burseraceae | Indian white mahogany | India | – Parenchyma rare (scanty) or absent (Fig. 10)<br>– Intervessel pits alternate and large<br>– Vessel-ray pits with much reduced borders to apparently simple: pits rounded or angular and horizontal (scalariform, gash-like) to vertical (palisade) (Fig. 11)<br>– No prismatic crystals<br>– Silica bodies in fibres |
| *Ceratopetalum apetalum* D.Don | Cunoniaceae | Rose mahogany<br>Coachwood | Australia | – Apotracheal parenchyma diffuse-in-aggregates in indistinct narrow bands (Fig. 12)<br>– No paratracheal parenchyma<br>– Simple and scalariform perforation plates<br>– Intervessel pits scalariform and opposite (Fig. 13)<br>– Prismatic crystals in ray cells and axial parenchyma (in chains) |

| Species | Family | Common name(s) | Geographical distribution | Distinctive wood anatomy |
|---|---|---|---|---|
| *Shorea* Roxb. ex Gaertner | Dipterocarpaceae | Philippine mahogany<br>Bataan mahogany<br>Lauan<br>Meranti | South-East Asia | – Apotracheal parenchyma diffuse-in-aggregates<br>– Paratracheal parenchyma scanty, vasicentric, lozenge-aliform, winged-aliform and confluent<br>– Axial intercellular canals in long tangential lines (Fig. 14)<br>– Rays commonly of two distinct sizes (Fig. 15)<br>– Intervessel pits medium to large, alternate, opposite and vestured<br>– Vessel-ray pits with much reduced borders to apparently simple: pits rounded or angular and horizontal (scalariform, gash-like) to vertical (palisade)<br>– Prismatic crystals in axial parenchyma: in chambered cells (including enlarged cells/idioblasts) and non-chambered cells |
| *Afzelia quanzensis* Welw. | Leguminosae-Caesalpinioideae | Pod-mahogany<br>Mahogany bean<br>Chanfuta | Africa | – Paratracheal parenchyma lozenge-aliform, winged-aliform and confluent (Fig. 16)<br>– Homocellular rays<br>– Intervessel pits small, medium, large<br>– Vestured pits<br>– Prismatic crystals in chambered axial parenchyma cells |
| *Gymnocladus dioica* L. | Leguminosae-Caesalpinioideae | American mahogany<br>Kentucky coffee tree | North America | – Wood ring-porous (Fig. 17)<br>– Vessels in tangential bands, in a diagonal/radial pattern (Fig. 17), latewood vessels in clusters<br>– Paratracheal parenchyma vasicentric, lozenge-aliform, winged-aliform and confluent<br>– Homocellular rays<br>– Vestured pits<br>– Helical thickenings in vessel elements<br>– No prismatic crystals |
| *Acacia koa* A.Gray | Leguminosae-Mimosoideae | Hawaiian mahogany<br>Koa | Hawaii | – Vessels of two diameter classes, not ring-porous (Fig. 18)<br>– Paratracheal parenchyma vasicentric<br>– Homocellular rays<br>– Vestured pits<br>– Prismatic crystals in chambered axial parenchyma cells (Fig. 19, see arrow) |
| *Plathymenia reticulata* Benth. | Leguminosae-Mimosoideae | Brazilian mahogany<br>Vinhatico | Brazil | – Paratracheal parenchyma vasicentric and confluent (Fig. 20)<br>– Homocellular rays<br>– Intervessel pits medium, alternate<br>– Vestured pits<br>– Prismatic crystals in chambered axial parenchyma cells, in chains |
| *Myroxylon balsamum* L. (Harms) | Leguminosae-Papilionoideae | Santos mahogany<br>Balsamo | Central & South America | – Paratracheal parenchyma scanty, vasicentric and confluent (Fig. 21)<br>– Storied rays, axial parenchyma and vessel elements (Fig. 22)<br>– Intervessel pits minute, small and alternate<br>– Vestured pits<br>– Prismatic crystals in ray cells and in chambered axial parenchyma |
| *Pterocarpus dalbergioides* Roxb. ex DC. | Leguminosae-Papilionoideae | East India mahogany | India | – Wood semi-ring and diffuse porous (Fig. 23)<br>– Vessels of two diameter classes, not ring-porous (Fig. 23)<br>– Apotracheal parenchyma diffuse and diffuse-in-aggregates (Fig. 23)<br>– Paratracheal parenchyma vasicentric, lozenge-aliform, winged-aliform and confluent (Fig. 23)<br>– Rays, axial parenchyma, vessel elements and fibres storied (Fig. 24)<br>– Rays uniseriate (Fig. 24) and heterocellular<br>– Intervessel pits small, medium, large<br>– Vestured pits<br>– Prismatic crystals present in chambered axial parenchyma cells, in chains |
| *Eucalyptus acmenioides* Schauer | Myrtaceae | White mahogany | Australia | – Vessels in a diagonal/radial pattern and exclusively solitary (Fig. 25)<br>– Apotracheal parenchyma diffuse (Fig. 25)<br>– Paratracheal parenchyma vasicentric (Fig. 25)<br>– Intervessel pits medium and alternate<br>– Vestured pits<br>– Vessel-ray pits with much reduced borders to apparently simple: pits rounded or angular and horizontal (scalariform, gash-like) vertical (palisade) |

**Table 2 (cont.)**

| Species | Family | Common name(s) | Geographical distribution | Distinctive wood anatomy |
|---|---|---|---|---|
| *Eucalyptus botryoides* Sm. | Myrtaceae | Bastard mahogany | Australia | – Vessels in a diagonal/radial pattern and exclusively solitary (Fig. 26)<br>– Tyloses (Fig. 26)<br>– Apotracheal parenchyma diffuse (Fig. 26)<br>– Paratracheal parenchyma vasicentric<br>– Intervessel pits medium and alternate (Fig. 27)<br>– Vestured pits<br>– Vessel-ray pits with much reduced borders to apparently simple: pits rounded or angular and horizontal (scalariform, gash-like) to vertical (palisade) |
| *Eucalyptus marginata* Sm. | Myrtaceae | Western Australia mahogany<br>Jarrah | Australia | – Vessels in a diagonal/radial pattern and exclusively solitary (Fig. 28)<br>– Tyloses (Figs. 28 and 29)<br>– Paratracheal parenchyma vasicentric<br>– Intervessel pits small, medium, large and alternate<br>– Vestured pits<br>– Vessel-ray pits with much reduced borders to apparently simple: pits rounded or angular and horizontal (scalariform, gash-like) to vertical (palisade) |
| *Cercocarpus* species Kunth. | Rosaceae | Mountain-mahoganies<br>Hardtack | North America | – Wood semi-ring and diffuse porous (Fig. 30)<br>– Vessels exclusively solitary<br>– Parenchyma absent or rare (diffuse or diffuse-in-aggregates)<br>– Simple, reticulate, foraminate and/or other types of perforation plates<br>– Helical thickenings in vessel elements and ground tissue fibres<br>– Prismatic crystals in axial parenchyma, in enlarged cells (idioblasts) |
| *Tieghemella heckelii* Pierre ex A. Chev. | Sapotaceae | Cherry mahogany<br>Makoré | Africa | – Vessels in a diagonal/radial pattern (Fig. 31)<br>– Long radial multiples<br>– Tyloses<br>– Parenchyma in narrow bands and reticulate (Fig. 31)<br>– Rays uniseriate and multiseriate, heterocellular (Fig. 32)<br>– Intervessel pits alternate, small and medium<br>– Vessel-ray pits with distinct borders similar to intervessel pits and also with much reduced borders to apparently simple: pits rounded or angular and horizontal (scalariform, gash-like) to vertical (palisade)<br>– Silica bodies in ray cells (Fig. 32) |
| *Heritiera utilis* Sprague (Sprague) | Sterculiaceae | Niangon<br>Cola mahogany | West Africa | – Large vessels present (Fig. 33)<br>– Paratracheal parenchyma diffuse and diffuse-in-aggregates and vasicentric (Fig. 33)<br>– Rays of two distinct sizes (Fig. 34)<br>– Sheath cells present (Fig. 34) |
| *Pentace burmanica* Kurz | Tiliaceae | Burma mahogany<br>Thitka | India | – Tyloses<br>– Apotracheal parenchyma diffuse and diffuse-in-aggregates (Fig. 35)<br>– Paratracheal parenchyma vasicentric and scalariform<br>– Rays, axial parenchyma, vessel elements and fibres storied (Fig. 36)<br>– Vessel-ray pits with distinct borders similar to intervessel pits and also with much reduced borders to apparently simple: pits rounded or angular, restricted to marginal rows |

# Non–Meliaceae mahoganies

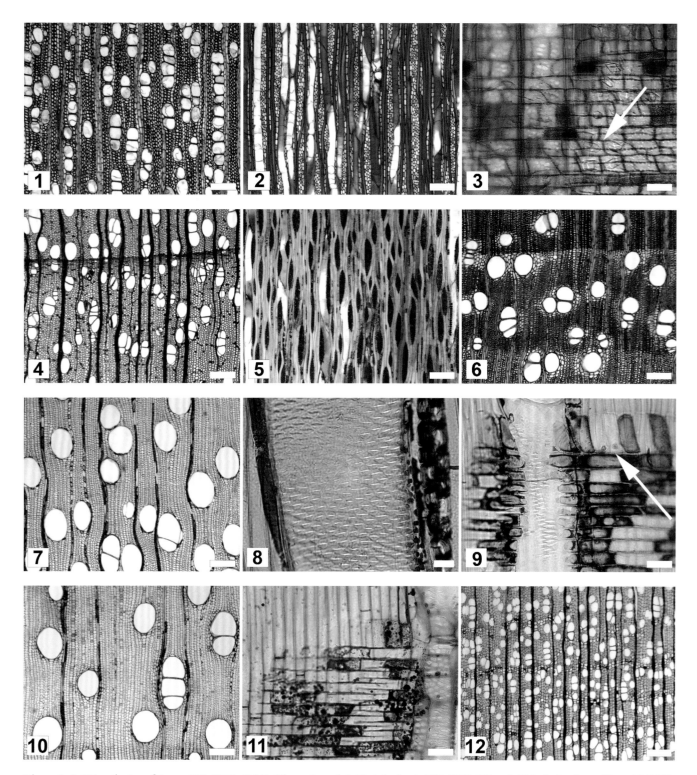

**Figs. 1–3** *Kiggelaria africana* TS, TLS, RLS. **Figs. 4 and 5** *Betula lenta* TS, TLS. **Fig. 6** *Tabebuia donnell-smithii* TS. **Figs. 7–9** *Aucoumea klaineana* TS, TLS, RLS. **Figs. 10 and 11** *Canarium euphyllum* TS, RLS. **Fig. 12** *Ceratopetalum apetalum* TS. Figs. 1, 2, 4–7, 10 & 12, scale bar = 200 μm. Figs. 3, 9 and 11, scale bar = 50 um. Fig. 8, scale bar = 20 μm. Arrow indicates presence of prismatic crystals in Fig. 3, and silica bodies in Fig. 9.

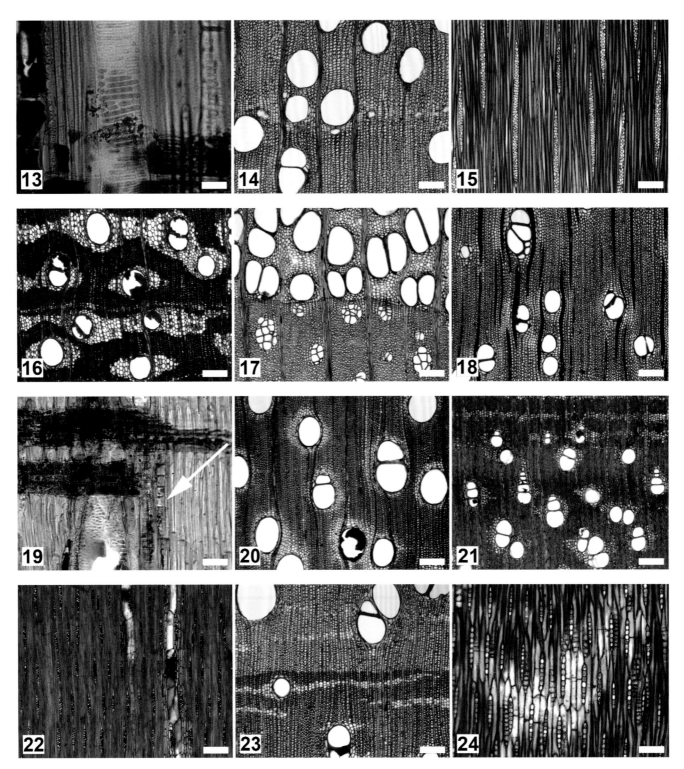

**Fig. 13** *Ceratopetalum apetalum* RLS. **Figs. 14 and 15** *Shorea polysperma* TS, TLS. **Fig. 16** *Afzelia quanzensis* TS. **Fig. 17** *Gymnocladus dioica* TS. **Figs. 18 and 19** *Acacia koa* TS, RLS. **Fig. 20** *Plathymenia reticulata* TS. **Figs. 21 and 22** *Myroxylon balsamum* TS, TLS. **Figs. 23 and 24** *Pterocarpus dalbergioides* TS, TLS. Figs. 13, 14, 16–18, 20–24, scale bar = 200 μm. Figs. 15 and 19, scale bar = 50 μm. Arrow indicates prismatic crystal in Fig. 19.

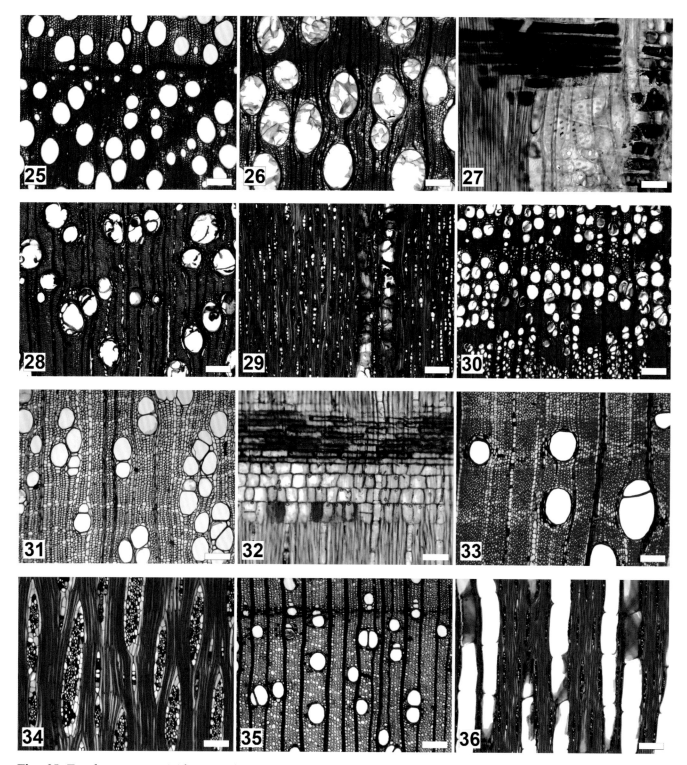

**Fig. 25** *Eucalyptus acmenioides* TS. **Figs. 26 and 27** *Eucalyptus botryoides* TS, RLS. **Figs. 28 and 29** *Eucalyptus marginata* TS, TLS. **Fig. 30** *Cercocarpus parviflorus* TS. **Figs. 31 and 32** *Tieghemella heckelii* TS, RLS. **Figs. 33 and 34** *Heritiera utilis* TS, TLS. **Figs. 35 and 36** *Pentace burmanica* TS, TLS. Figs. 25, 26, 28–31, & 33–36, scale bar = 200 μm. Fig. 27, scale bar = 50 μm. Fig. 32, scale bar = 100 μm.

# APPENDIX 1

## Details of Meliaceae microscope slides examined

For details of wood collections at Kew, Madison and elsewhere see Stern (1988).

| Species | Author | RBG Kew slide database number | Wood collection catalogue number | Slide label information |
|---|---|---|---|---|
| *Aglaia argentea* | Blume | 21944 | | Java |
| *Aglaia argentea* | Blume | 21945 | | Bot. Gdns., Singapore |
| *Aglaia argentea* | Blume | 21946 | | Burma |
| *Aphanamixis polystachya* | (Wall.) R.Parker | 21960 | | Fairchild Trop. Gdn. 6993 A SF50 |
| *Azadirachta excelsa* | (Jack) Jacobs | 21961 | | Leiden BW 4471 MPFSL |
| *Azadirachta indica* | A.Juss. | 21963 | | 1974 MPFSL |
| *Azadirachta indica* | A.Juss. | 7192 | | Nigeria 1992–1833 |
| *Cabralea canjerana* | (Vell.) Mart. | 21964 | | Brazil 16583 FPRL |
| *Carapa guianensis* | Aubl. | 24558 | Kw 70051 | Trinidad & Tobago "Crappo" |
| *Carapa guianensis* | Aubl. | 21972 | | Trinidad |
| *Carapa guianensis* | Aubl. | 21973 | | Trinidad 54–1920 |
| *Carapa guianensis* | Aubl. | 21974 | | CNHM 623341 Madison |
| *Carapa procera* | DC. | 21980 | Kw 3462 | Ghana |
| *Cedrela fissilis* | Vell. | 21985 | | Paraguay |
| *Cedrela fissilis* | Vell. | 21986 | | Argentina |
| *Cedrela fissilis* | Vell. | 21987 | | Argentina |
| *Cedrela fissilis* | Vell. | 21988 | | W12 USNM Div. Woods |
| *Cedrela fissilis* | Vell. | 21989 | | Sao Paulo SPFw 220 |
| *Cedrela fissilis* | Vell. | 21990 | | Madison MPFSL Y.N. 10533 |
| *Cedrela odorata* | L. | 21402 | Kw 3989 | Surinam G. Stahel |
| *Cedrela odorata* | L. | 21403 | Kw 3993 | Brazil |
| *Cedrela odorata* | L. | 21404 | Kw 3996 | |
| *Cedrela odorata* | L. | 21999 | | Br. Honduras |
| *Cedrela odorata* | L. | 22001 | | FPRL |
| *Cedrela odorata* | L. | 21401 | Kw 4000 | Singapore |
| *Cedrela odorata* | L. | 21996 | MADw 21261 | |
| *Cedrela odorata* | L. | 21998 | | |
| *Cedrela odorata* | L. | 21997 | | Dominica 12–1938 |
| *Chisocheton divergens* | Blume | | Kw 4031 | Malaysia 151–1894 HN Ridley Esq. |
| *Chisocheton divergens* | Blume | | Kw 4030 | Burma 8213 27. 1932 |
| *Chisocheton divergens* | Blume | 22016 | | Burma IFI 5351 |
| *Chisocheton divergens* | Blume | 22017 | | Burma E 8213 |
| *Chukrasia tabularis* | A. Juss. | 22019 | | Ind. For. Dept. 1878 |
| *Chukrasia tabularis* | A. Juss. | 22020 | | Assam Gamble 76 |
| *Chukrasia tabularis* | A. Juss. | 22021 | | Burma P. R. 7411 |
| *Chukrasia tabularis* | A. Juss. | 22022 | | FPRL 2/939 |
| *Chukrasia tabularis* | A. Juss. | 22023 | | Burma |
| *Dysoxylum fraserianum* | Benth. | 22031 | | Australia |
| *Dysoxylum fraserianum* | Benth. | 22032 | | Australia 52–1936 |
| *Dysoxylum fraserianum* | Benth. | 22033 | | Australia FPRL U. 7-6 |
| *Entandrophragma angolense* | C.DC. | 22079 | | Uganda IFI 8298 |
| *Entandrophragma angolense* | C.DC. | 22078 | | Cameroon Vic. Bot. Gdns. 18 |

(cont.)

| Species | Author | RBG Kew slide database number | Wood collection catalogue number | Slide label information |
|---|---|---|---|---|
| *Entandrophragma candollei* | Harms | 22080 | | Nigeria FPRL 46 1932 |
| *Entandrophragma cylindricum* | Sprague | 24703 | Kw 4170 | |
| *Entandrophragma cylindricum* | Sprague | 22082 | | Gold Coast |
| *Entandrophragma cylindricum* | Sprague | 22087 | | Africa 29-1931 |
| *Entandrophragma cylindricum* | Sprague | 22084 | | 5115 |
| *Entandrophragma cylindricum* | Sprague | 22085 | | W. Africa |
| *Entandrophragma cylindricum* | Sprague | 22083 | | Gold Coast 1946 |
| *Entandrophragma cylindricum* | Sprague | 22089 | | W. Africa John Wright & Sons 1936 |
| *Entandrophragma cylindricum* | Sprague | 22086 | | Uganda "Sapele"/ "Muyovu" |
| *Entandrophragma cylindricum* | Sprague | 22090 | | Gold Coast FPRL 29-1931 |
| *Entandrophragma cylindricum* | Sprague | 22088 | | Cameroon Vict. Bot. Gdns. 19 |
| *Entandrophragma utile* | Sprague | 22004 | | Uganda CFI 8154 |
| *Entandrophragma utile* | Sprague | | | Uganda |
| *Entandrophragma utile* | Sprague | | | Uganda |
| *Entandrophragma utile* | Sprague | 22095 | | Uganda CFI 8148 |
| *Entandrophragma utile* | Sprague | 22096 | | Sierra Leone 48 1924 |
| *Entandrophragma utile* | Sprague | 22097 | | Tervuren Mus. 1410 |
| *Guarea cedrata* | Pellegr. ex Chev. | 22107 | | CNHM 622804 Madison |
| *Guarea cedrata* | Pellegr. ex Chev. | 22108 | | W. Africa 1973 |
| *Guarea cedrata* | Pellegr. ex Chev. | 22106 | | Gold Coast 1945 |
| *Guarea thompsonii* | Sprague & Hutch. | 22118 | | Tervuren No. 1110 |
| *Guarea thompsonii* | Sprague & Hutch. | 22119 | | Dept. For. Nigeria |
| *Guarea thompsonii* | Sprague & Hutch. | 22120 | | Gold Coast FRR |
| *Guarea thompsonii* | Sprague & Hutch. | 22121 | | West Africa Trade 1862 |
| *Guarea thompsonii* | Sprague & Hutch. | 22122 | | West Africa MPFSL 1973 |
| *Guarea thompsonii* | Sprague & Hutch. | 22123 | | Madison MPFSL 1974 |
| *Khaya anthotheca* | C.DC. | 24705 | Kw 4214 | Ghana |
| *Khaya anthotheca* | C.DC. | 22126 | | FPRL |
| *Khaya anthotheca* | C.DC. | 22127 | | CNHM 622778 Madison |
| *Khaya grandifoliola* | C.DC. | 24706 | Kw 4220 | Nigeria |
| *Khaya grandifoliola* | C.DC. | 22128 | | |
| *Khaya grandifoliola* | C.DC. | 22128 | | W. Africa Gold Coast No. 1803 |
| *Khaya grandifoliola* | C.DC. | 22129 | | |
| *Khaya ivorensis* | A.Chev. | 24702 | Kw 4224 | Nigeria (Gold Coast) |
| *Khaya ivorensis* | A.Chev. | 22130 | | |
| *Khaya ivorensis* | A.Chev. | 22131 | | |
| *Khaya ivorensis* | A.Chev. | 22134 | | |
| *Khaya ivorensis* | A.Chev. | 22135 | | |
| *Khaya ivorensis* | A.Chev. | 22136 | | |
| *Khaya ivorensis* | A.Chev. | 22137 | | |
| *Khaya ivorensis* | A.Chev. | 22138 | | CNHM 622781 Madison |
| *Khaya ivorensis* | A.Chev. | 22139 | | CNHM 614232 Madison |
| *Khaya nyasica* | Stapf | 24707 | Kw 4234 | Mozambique |
| *Khaya nyasica* | Stapf | 22140 | | Katanga "Mululu" |
| *Khaya nyasica* | Stapf | 22141 | | Mozambique 1-1913 |
| *Khaya senegalensis* | A.Juss. | 24704 | Kw 71646 | Gold Coast |
| *Khaya senegalensis* | A.Juss. | 22142 | Kw 4223 | Nigeria |
| *Khaya senegalensis* | A.Juss. | 22143 | | |
| *Khaya senegalensis* | A.Juss. | 22144 | | Nyasaland FPRL 2012 |
| *Khaya senegalensis* | A.Juss. | 22145 | | W. Africa Gold Coast No. 1569 |
| *Lovoa trichilioides* | Harms | 22151 | | Africa FPRL 474 |
| *Lovoa trichilioides* | Harms | 22152 | | Trade Sample 1964 |
| *Lovoa trichilioides* | Harms | 22154 | | W. Africa FPRL 3176 |
| *Lovoa trichilioides* | Harms | 22156 | | W. Africa FRR Martin |
| *Lovoa trichilioides* | Harms | 22155 | | W. Africa Martin 1948 |
| *Melia azedarach* | L. | 22158 | | Brazil CEPEC 203 |

(cont.)

| Species | Author | RBG Kew slide database number | Wood collection catalogue number | Slide label information |
|---|---|---|---|---|
| *Melia azedarach* | L. | 22159 | K656 | USA |
| *Melia azedarach* | L. | 22160 | BWCw 8315 | USA |
| *Melia azedarach* | L. | 22161 | BWCw 8368 | USA |
| *Melia azedarach* | L. | 22162 | K8594 | USA |
| *Melia azedarach* | L. | 22163 | | Gold Coast IFI 4291 |
| *Melia azedarach* | L. | 22164 | | Japan Mus. IV |
| *Melia composita* | Willd. | 22168 | MADw 61202 | Coll 762 |
| *Melia composita* | Willd. | 22169 | | Ceylon |
| *Melia composita* | Willd. | 22176 | | |
| *Pseudocedrela kotschyi* | Harms | 22174 | | W. Africa Trade 1960 |
| *Swietenia humilis* | Zucc. | 28560 | BRE 27505 | Nicaragua FPRL 53/76/16 |
| *Swietenia humilis* | Zucc. | 28837 | SJRw 7483 | Salvador |
| *Swietenia humilis* | Zucc. | 28838 | SJRw 7484 | Salvador Dr S Calderon 1924 |
| *Swietenia humilis* | Zucc. | 28890 | SJRw 4765 | Oaxaca, Mexico Flora Neotropica 28:442 |
| *Swietenia humilis* | Zucc. | 29035 | SJRw 8902 | Puerto Barrios, Guatemala |
| *Swietenia humilis* | Zucc. | 29036 | SJRw 38312 | Nicaragua Det. by Record & Hess |
| *Swietenia macrophylla* | King | 24750 | Kw 4357 | India |
| *Swietenia macrophylla* | King | 24701 | Kw 4353 | Mexico (Chiapas) |
| *Swietenia macrophylla* | King | 24696 | Kw 4348 | Sri Lanka "Large leaf mahogany" |
| *Swietenia macrophylla* | King | 22189 | | Peru |
| *Swietenia macrophylla* | King | 22190 | | 3964 |
| *Swietenia macrophylla* | King | 22191 | | Fiji 20471 |
| *Swietenia macrophylla* | King | 22192 | | Fiji S. 1401-4 |
| *Swietenia macrophylla* | King | 22193 | | Br. Honduras FRR |
| *Swietenia macrophylla* | King | 22194 | | Brazil CEPECw 176 |
| *Swietenia macrophylla* | King | 22195 | | CNHM 623433 Madison |
| *Swietenia mahagoni* | L.Jacq. | 24710 | Kw 4370 | Cuba |
| *Swietenia mahagoni* | L.Jacq. | 24559 | Kw 4376 | West Indies "American /Spanish mahogany" |
| *Swietenia mahagoni* | L.Jacq. | 24709 | Kw 4388 | Sri Lanka "Small leaf mahogany" |
| *Swietenia mahagoni* | L.Jacq. | 22196 | | Florida |
| *Swietenia mahagoni* | L.Jacq. | 22197 | | West Indies "Cuban mahogany" |
| *Swietenia mahagoni* | L.Jacq. | 22200 | | Florida 60-1883 |
| *Swietenia mahagoni* | L.Jacq. | 22201 | | West Indies 1921 |
| *Swietenia mahagoni* | L.Jacq. | 22202 | | |
| *Swietenia mahagoni* | L.Jacq. | 22203 | | Florida |
| *Swietenia mahagoni* | L.Jacq. | 22198 | | Calcutta Bot. Gdn. |
| *Swietenia mahagoni* | L.Jacq. | 22199 | | Jamaica 73 |
| *Swietenia mahagoni* | L.Jacq. | 21400 | Kw 4371 | Jamaica |
| *Swietenia mahagoni* | L.Jacq. | 21398 | Kw 4379 | India F. H. Pierpoint |
| *Swietenia mahagoni* | L.Jacq. | 21399 | Kw 4380 | F. H. Pierpoint |
| *Swietenia mahagoni* | L.Jacq. | 21397 | Kw 4374 | USA |
| *Toona ciliata* | M.Roem | | Kw 4407 | Singapore Burkill 2315 16.1923 |
| *Toona ciliata* | M.Roem | | Kw 4401 | India Ind. For. Dept. 1878 |
| *Toona ciliata* | M.Roem | 22005 | | India Trade 1948 |
| *Toona ciliata* | M.Roem | 22006 | | India Gamble's Coll. |
| *Toona ciliata* | M.Roem | 22007 | | Burma Gamble's Coll. |
| *Toona ciliata* | M.Roem | 22008 | | Siam Roy. For. Dept. 17-1914 |
| *Toona ciliata* | M.Roem | 22009 | | Singapore Burkhill 8. 1916 |
| *Toona ciliata* | M.Roem | 22010 | | Darjeeling Ind. For. Dept. 1878 |
| *Toona ciliata* | M.Roem | 22011 | | Singapore Bot. Gdns. 449 |
| *Toona ciliata* | M.Roem | 22012 | | Burma E9354 |
| *Toona ciliata* | M.Roem | 22013 | | Australia |
| *Turraeanthus africanus* | (Welw.) Pellegr. | 22131 | | W. Africa 1938 |
| *Turraeanthus africanus* | (Welw.) Pellegr. | 22230 | | Cameroon Victoria Bot. Gdn. |
| *Turraeanthus africanus* | (Welw.) Pellegr. | 22221 | | Gold Coast Pellegrin 1945 |
| *Turraeanthus africanus* | (Welw.) Pellegr. | 22232 | Kw 24729 | |

# APPENDIX 2

## Trade data for *Swietenia* 1996–2005

The gross import and gross export volumes are given for *Swietenia* from 1996 to 2005 (the latest full year of recorded trade data). The top importing and exporting countries are included; those with less than 0.1 % have been excluded from the table. All units are in cubic metres, and a conversion factor of 600kg/m$^3$ has been used where quantity is recorded in weight. The parts are listed as veneer, sawn wood or timber. Data that could not be converted to cubic metres or had no units at all have been omitted, as the amount of material imported or exported could not be established. CITES trade statistics derived from the CITES trade database, UNEP World Conservation Monitoring Centre, Cambridge, U.K.

### *Swietenia* gross import (units m$^3$)

| Country | 1996 | 1997 | 1998 | 1999 | 2000 | 2001 | 2002 | 2003 | 2004 | 2005 | Total 1996–2005 | % of total import |
|---|---|---|---|---|---|---|---|---|---|---|---|---|
| USA | 56710 | 91375 | 77292 | 92808 | 75131 | 89040 | 82248 | 60518 | 43608 | 52567 | 721297 | 55.6 |
| Dominican Republic | 10426 | 11336 | 5606 | 19391 | 14919 | 10260 | 18200 | 1820 | 306320 | 4526 | 402804 | 31.1 |
| Canada | 10 | 28 | 11199 | 3835 | 3538 | 2545 | 24632 | 1766 | 1708 | 1787 | 51048 | 3.9 |
| UK | 16931 | 1825 | 4166 | 5682 | 2813 | 2981 | 1433 | 61 | 381 | 231 | 36504 | 2.8 |
| Mexico | 915 | 194 | 333 | 205 | 622 | 2675 | 2979 | 2447 | 3573 | 1131 | 15074 | 1.2 |
| Spain | 791 | 824 | 3976 | 2198 | 969 | 815 | 886 | 209 | 210 | 440 | 11318 | 0.9 |
| The Netherlands | 890 | 537 | 2198 | 2824 | 1183 | 600 | 737 | 41 | 0 | 0 | 9010 | 0.7 |
| Japan | 252 | 0 | 37 | 94 | 27 | 26 | 1 | 91 | 251 | 6821 | 7600 | 0.6 |
| Denmark | 1764 | 753 | 237 | 524 | 548 | 627 | 279 | 381 | 212 | 61 | 5386 | 0.4 |
| Germany | 29 | 254 | 871 | 585 | 332 | 579 | 434 | 96 | 345 | 638 | 4163 | 0.3 |
| Ireland | 2302 | 1146 | 310 | 145 | 84 | 16 | 18 | 18 | 16 | 0 | 4055 | 0.3 |
| Puerto Rico | 124 | 326 | 105 | 736 | 694 | 452 | 204 | 413 | 379 | 261 | 3694 | 0.3 |
| Cuba | 634 | 832 | 458 | 224 | 30 | 0 | 0 | 81 | 0 | 0 | 2259 | 0.2 |
| Hong Kong | 0 | 0 | 2004 | 9 | 0 | 32 | 0 | 10 | 0 | 0 | 2055 | 0.2 |
| Argentina | 249 | 219 | 1031 | 340 | 113 | 34 | 0 | 0 | 0 | 46 | 2032 | 0.2 |
| Belgium | 122 | 580 | 393 | 87 | 212 | 203 | 5 | 5 | 152 | 56 | 1815 | 0.1 |
| Honduras | 167 | 143 | 0 | 0 | 393 | 0 | 446 | 0 | 265 | 301 | 1715 | 0.1 |
| Cayman Islands | 0 | 0 | 0 | 1337 | 0 | 0 | 0 | 0 | 0 | 0 | 1337 | 0.1 |
| Chile | 85 | 42 | 76 | 264 | 300 | 83 | 220 | 59 | 130 | 68 | 1327 | 0.1 |
| Italy | 32 | 0 | 140 | 348 | 132 | 202 | 287 | 5 | 2 | 0 | 1148 | 0.1 |
| Sweden | 0 | 114 | 17 | 272 | 71 | 143 | 219 | 79 | 25 | 205 | 1145 | 0.1 |
| Australia | 4 | 0 | 175 | 303 | 132 | 80 | 93 | 237 | 0 | 3 | 1027 | 0.1 |
| Portugal | 0 | 0 | 315 | 2 | 0 | 0 | 595 | 0 | 0 | 0 | 912 | 0.1 |
| France | 403 | 200 | 31 | 32 | 3 | 0 | 66 | 4 | 37 | 29 | 805 | 0.1 |
| Norway | 130 | 98 | 99 | 210 | 0 | 76 | 79 | 0 | 0 | 33 | 725 | 0.1 |
| Dominica | 705 | 0 | 0 | 0 | 0 | 0 | 0 | 0 | 0 | 0 | 705 | 0.1 |
| China | 0 | 0 | 0 | 0 | 147 | 63 | 118 | 31 | 39 | 264 | 662 | 0.1 |

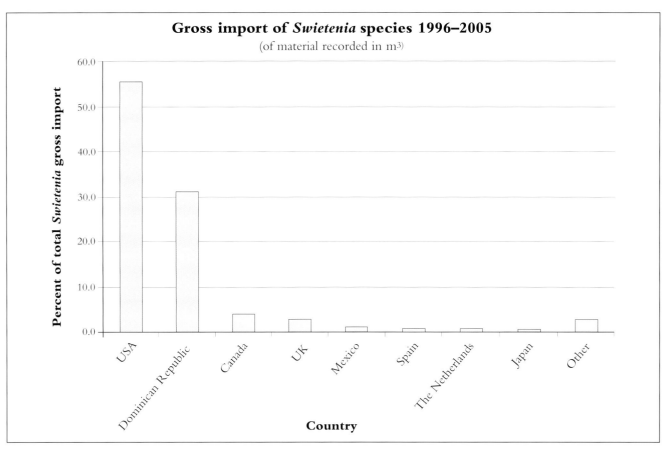

**Gross import of *Swietenia* species 1996–2005**

(of material recorded in m³)

**Gross export of *Swietenia* species 1996–2005**

(of material recorded in m³)

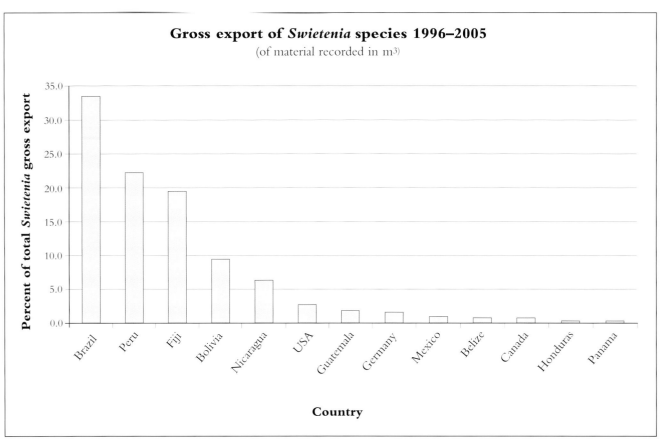

## *Swietenia* gross export (units m³)

| Country | 1996 | 1997 | 1998 | 1999 | 2000 | 2001 | 2002 | 2003 | 2004 | 2005 | Total 1996–2005 | % of total export |
|---|---|---|---|---|---|---|---|---|---|---|---|---|
| Brazil | 101969 | 111872 | 48627 | 60052 | 43491 | 44880 | 44952 | 8876 | 4750 | 57 | 469526 | 33.0 |
| Peru | 5932 | 10893 | 20775 | 51549 | 33049 | 41499 | 54043 | 42672 | 31231 | 24926 | 316569 | 22.3 |
| Fiji | 0 | 0 | 0 | 0 | 0 | 0 | 0 | 0 | 276932 | 0 | 276932 | 19.5 |
| Bolivia | 25989 | 28048 | 20159 | 8519 | 10549 | 7612 | 7182 | 8603 | 9743 | 7957 | 134361 | 9.5 |
| Nicaragua | 17105 | 19127 | 5983 | 5164 | 3978 | 5991 | 7278 | 158 | 22786 | 4004 | 91574 | 6.4 |
| USA | 506 | 833 | 11417 | 4200 | 3899 | 2480 | 6196 | 591 | 4583 | 3263 | 37968 | 2.7 |
| Guatemala | 2100 | 1700 | 1098 | 439 | 2716 | 3134 | 2482 | 971 | 5812 | 6466 | 26918 | 1.9 |
| Germany | 0 | 0 | 0 | 0 | 4 | 37 | 11 | 3 | 20 | 21324 | 21399 | 1.5 |
| Mexico | 2590 | 927 | 271 | 211 | 44 | 2473 | 4653 | 300 | 482 | 688 | 12639 | 0.9 |
| Belize | 1931 | 233 | 125 | 2326 | 2029 | 709 | 1239 | 1859 | 1111 | 434 | 11996 | 0.8 |
| Canada | 0 | 0 | 2250 | 454 | 1462 | 223 | 5498 | 1121 | 57 | 27 | 11092 | 0.8 |
| Honduras | 0 | 884 | 907 | 1324 | 665 | 570 | 0 | 39 | 282 | 133 | 4804 | 0.3 |
| Panama | 0 | 0 | 71 | 23 | 1202 | 2135 | 765 | 56 | 0 | 0 | 4252 | 0.3 |

# FURTHER READING

Acevedo Mallque, M. & Kikata, Y. (1994). *Atlas de maderas del Peru*. National Agrarian University, La Molina, Peru.

Barajas-Morales, J. & Gomez, C. L. (1989). *Anatomia de maderas de Mexico: especies de una selva baja caducifolia*. Universidad Nacional Autónoma de México.

Barros, C. F. & Callado, C. H. (1997). *Madeiras da Mata Atlantica: anatomia do lenho de especies ocorrentes nos remanescentes florestais do estado do Rio de Janeiro, Brasil*. Vol. 1. [*Timbers of the Atlantic rain forest: wood anatomy of species from remnant forests in Rio de Janeiro state, Brazil*. Vol. 1]. Jardim botânico do Rio de Janeiro, Brazil.

Beeckman, H. (2003). De microscopische schoonheid van mahonie. *Interdisciplinair tijdschrift voor Conservering en Restauratie* 4: 18–27.

Blundell, A. G. (2004). A review of the CITES listing of big-leaf mahogany. *Oryx* 38: 84–90.

Boerhave Beekman, W. (1964). *Elsevier's Wood Dictionary*. Vol. 1: *Commercial and botanical nomenclature of world timbers; sources of supply*. Elsevier Pub. Co., Amsterdam, New York.

Brazier, J. D. & Franklin, G. L. (1961). *Identification of hardwoods: a microscope key*. Her Majesty's Stationery Office, London.

British Standards Institution (1991). *British standard nomenclature of commercial timbers including sources of supply* BS 7359:1991. BSI, London.

Burkill, H. M. (1997). Meliaceae. In. *The useful plants of West Tropical Africa. Vol. 4 Families M–R*, pp. 88–134, Royal Botanic Gardens, Kew.

Camargos, J. A. A., Coradin, V. T. R., Czarneski, C. M., Oliveira, D. de & Meguerditchian, I. (2001). *Catálago de arvores do Brasil*. IBAMA, Brasília.

Chalk, L., Burtt Davy, J., Desch, H. E. & Hoyle, A. C. (1933). *Twenty West African timber trees*. Clarendon Press, Oxford.

Chalmers, K. J., Newton, A. C., Waugh, R., Wilson, J. & Powell, W. (1994). Evaluation of the extent of genetic variation in mahoganies (Meliaceae) using RAPD markers. *Theoretical and Applied Genetics* 89: 504–508.

CITES (Convention on International Trade in Endangered Species of Wild Fauna and Fauna) website. www.cites.org

CITES trade database, UNEP World Conservation Monitoring Centre, Cambridge, U.K. www.unep-wcmc.org/citestrade/trade.cfm [2007]

Datta, P. C. & Samanta, P. (1983). Anatomy of some Indo-Malayan Meliaceae. *Journal of the Indian Botanical Society* 62: 185–203.

Desch, H. E. (1954). *Manual of Malaysian timbers*. Malaysian Forest Records No. 15. Vol. 2. pp. 334–356.

Détienne, P. & Jacquet, P. (1983). *Atlas d'identification des bois de l'Amazonie et des régions voisines*. Centre Technique Forestier Tropicale, Nogent-sur-Marne, France.

Détienne, P. Jacquet, P. & Mariaux, A. (1982). *Manuel d'identification des bois tropicaux. Tome 3: Guyane française*. Centre Technique Forestier Tropicale, Nogent-sur-Marne, France.

Dixon, H. H. (1919). Mahogany and the recognition of some of the different kinds by their microscopic characteristics. *Notes from the Botanical School of Trinity College, Dublin*. 3: 3–58.

Donaldson, L. A. (1984). Wood anatomy of five exotic hardwoods grown in Western Samoa. *New Zealand Journal of Forestry Science* 14: 305–318.

Dünisch, O., Bauch, J. & Gasparotto, L. (2002). Formation of increment zones and intraannual growth dynamics in the xylem of *Swietenia macrophylla*, *Carapa guianensis*, and *Cedrela odorata* (Meliaceae). *IAWA Journal* 23: 101–119.

Eggeling, W. J. & Harris, C. M. (1939). *Fifteen Uganda timbers*. Clarendon Press, Oxford.

Espinoza de Pernia, N. (1987). Estudio xilologico de algunas especies de *Cedrela* y *Toona*. *Pittieria* 14: 5–32.

Flynn, J. H. & Holder, C. D. (eds.). (2001). *A guide to useful woods of the world*. Forest Products Society Madison, Wisconsin.

Fouarge, J. & Gérard, G. (1964). *Bois du Mayumbe*. Publications de l'institut national pour l'étude agronomique du Congo, Bruxelles.

Fouarge, J., Gérard, G. & Sacré, E. (1953). *Bois du Congo*. Publications de l'institut national pour l'étude agronomique du Congo, Bruxelles.

Freitas, M. C. P. G. (1987). *Madeiras de S. Tomé: características anatómicas e físicas*. Centro de Estudos de Tecnologia Florestal do Instituto de Investigação Científica Tropical, Lisboa.

Gaiotti de Peralta, C. & Edlmann Abbate, M. L. (1981). Caratteristiche anatomiche ed usi di 25 specie legnose provenienti dalla Repubblica di Panama. *Rivta di Agricoltura Subtropicale e Tropicale* 75: 325–379.

Ghosh, S. S., Purkayastha, S. K. & Lal, K. (1963). Meliaceae. In: *Indian Woods: their identification, properties, and uses*. Volume 2, pp. 81–159. Manager of Publications, Delhi.

Gleason, H. A. & Panshin, A. J. (1936). *Swietenia krukovii*: a new species of mahogany from Brazil. *American Journal of Botany* 23: 21–26.

Groom, P. (1926). Excretory systems in the secondary xylem of Meliaceae. *Annals of Botany* 40: 631–649.

Gullison, R. E., Panfil, S. N., Strouse, J. J. & Hubbell, S. (1996). Ecology and management of mahogany (*Swietenia macrophylla* King) in the Chimanes Forest, Beni, Bolivia. *Botanical Journal of the Linnean Society* 122: 3–34.

InsideWood. (2004 – onwards). Published on the Internet. http://insidewood/lib.ncsu.edu/search [July 2006].

International Tropical Timber Organization (2004). Making the mahogany trade work: report of the workshop on capacity-building for the implementation of the CITES Appendix-II listing of mahogany. International Tropical Timber Organization, Yokohama.

IUCN (2006). 2006 IUCN Red List of Threatened Species. www.iucnredlist.org. Downloaded on 1 August 2006.

Koehler, A. (1922). The identification of true mahogany, certain so-called mahoganies, and some common substitutes. U.S. Department of Agriculture Bulletin 1050: pp. 1–18.

Kribs, D. A. (1930). Comparative anatomy of the woods of the Meliaceae. *American Journal of Botany* 17: 724–738.

Kribs, D. A. (1959). *Commercial foreign woods on the American market*. Buckhout Laboratory, Department of Botany, The Pennsylvania State University, University Park, Pennsylvania.

Lamb, F. B. (1966). *Mahogany of Tropical America: its ecology and management*. University of Michigan Press, Ann Arbor (Michigan).

Lebacq, L. (1963). *Atlas anatomique des bois du Congo*. Volume 5. Publications de l'institut national pour l'étude agronomique du Congo, Bruxelles.

Lebacq, L. & Istas, J. R. (1950). *Les Bois des Méliacées du Congo Belge*. Tervuren, Belgium.

Lemmens, R. H. M. J., Soerianegara, I. & Wong, W. C. (eds.). (1995). PROSEA (Plant Resources of South-East Asia) No. 5 (2), Timber trees: minor commercial timbers. Backhuys Publishers, Leiden.

Lincoln, W. A. (1986). *World woods in colour*. Stobart, London.

Mabberley, D. J. (1997). *The plant book*. Second edition. Cambridge University Press.

Mabberley, D. J., Pannell, C. M. & Singh, A. M. (1995). *Flora Malesiana*, Series 1, Spermatophyta: flowering plants. Volume 12, part 1, Meliaceae. Rijksherbarium/Hortus Botanicus, Leiden University, Netherlands.

Melville, R. (1936). A list of true and false mahoganies. *Bulletin of Miscellaneous Information, Royal Botanic Gardens, Kew* 3: 193–210.

Metcalfe, C. R. & Chalk, L. (1950). *Anatomy of the Dicotyledons*. Volume 1. Clarendon Press, Oxford.

Miller, N. G. (1990). The genera of Meliaceae in the Southeastern United States. *Journal of the Arnold Arboretum* 71: 453–486.

Moeller, J. (1876). Beiträge zur vergleichenden Anatomie des Holzes. Denkschriften der mathematischen -naturwissenschaftlichen Classe d. 36, Wien. pp. 297–426.

Muellner, A. N., Samuel, R., Johnson, S. A., Cheek, M., Pennington, T. D. & Chase, M. W. (2003). Molecular phylogenetics of Meliaceae (Sapindales) based on nuclear and plastid DNA sequences. *American Journal of Botany* 90: 471–480.

Nair, M. N. B. (1987). Occurrence of helical thickenings in the vessel element walls of dicotyledonous woods. *Annals of Botany* 60: 23–32.

Nair, M. N. B. (1988). Wood anatomy and heartwood formation in Neem (*Azadirachta indica* A. Juss.). *Botanical Journal of the Linnean Society* 97: 79–90.

Nair, M. N. B. (1991). Wood anatomy of some members of the Meliaceae. *Phytomorphology* 41: 63–73.

Negi, K., Gupta, S., Chauhan, L. & Pal, M. (2003). Patterns of crystal distribution in the woods of Meliaceae from India. *IAWA Journal* 24: 155–162.

Normand, D. (1955). *Atlas des bois de la Côte d'Ivoire*. Centre Technique Forestier Tropical, Nogent-sur-Marne, France.

Normand, D. & Paquis, J. (1976). *Manuel d'identification des bois commerciaux*. Tome 2: Afrique guinée-congolaise. Centre Technique Forestier Tropicale, Nogent-sur-Marne, France.

Panshin, A. J. (1933). Comparative anatomy of the woods of the Meliaceae, sub-family Swietenioideae. *American Journal of Botany* 20: 638–668.

Pearson, R. S. & Brown, H. P. (1932). *Commercial timbers of India*. Volume 1. Government of India Central Publication Branch, Calcutta.

Pennington, T. D. (2002). Mahogany: carving a future. *Biologist* 49: 204–208.

Pennington, T. D. & Styles, B. T. (1975). A generic monograph of the Meliaceae. *Blumea* 22: 419–540.

Pennington, T. D. & Styles, B. T. (1981). *Meliaceae*. Flora Neotropica monograph number 28. Published for Organization for Flora Neotropica by the New York Botanical Garden, New York.

Phongphaew, P. (2003). *The commercial woods of Africa: a descriptive full-colour guide*. Stobart- Davies, Ammanford.

Read, M. (1990). *Mahogany: forests or furniture?* Fauna and Flora Preservation Society, Brighton.

Record, S. J. (1924). Meliaceae. In: *Timbers of tropical America*, pp. 340–363. New Haven, Yale University Press.

Record, S. J. & Hess, R. W. (1943). Meliaceae. In: *Timbers of the New World*, pp. 359–375. New Haven, Yale University Press.

Rendle, B. J. (1938). *Commercial mahoganies and allied timbers.* Forest Products Research Bulletin No. 18. His Majesty's Stationery Office, London.

Rendle, B. J. (1969). *World timbers*, Volumes 1–3. Ernest Benn Limited, London.

Shah, J. P. (2004). Anatomical structure and components of secondary vascular tissues in *Azadirachta indica* A. Juss. *Geobios* 31: 173–176.

Snook, L. K. (1996). Catastrophic disturbance, logging and the ecology of mahogany (*Swietenia macrophylla* King): grounds for listing a major tropical timber species in CITES. *Botanical Journal of the Linnean Society* 122: 35–46.

Soerianegara, I. & Lemmens, R. H. M. J. (eds.). (1993). PROSEA (Plant Resources of South-East Asia) No. 5 (1), *Timber trees: major commercial timbers.* Pudoc Scientific Publishers, Wageningen.

Sosef, M. S. M., Hong, L. T. & Prawirohatmodjo, S. (eds.). (1998). PROSEA (Plant Resources of South-East Asia) No. 5 (3), *Timber trees: lesser-known timbers.* Backhuys Publishers, Leiden.

Stern, W. L. (1988). Index Xylariorum: institutional wood collections of the world. 3rd revised edition. *IAWA Bulletin* 9: 203–252.

Stevens, P. F. (2001 onwards). Angiosperm Phylogeny Website. Version 7, May 2006. http://www.mobot.org/MOBOT/research/APweb/.

Styles, B. T. & White, F. (1991). *Flora of tropical East Africa: Meliaceae.* A. A. Balkema, Rotterdam.

Wagenführ, R. & Steiger, A. (1963). Mahagoni und mahagoniähnliche Hölzer-Probleme ihrer Identifizierung. *Holztechnologie* 4: 137–146.

Walker, A. (ed.). (2005). *The Encyclopaedia of Wood.* Quatro Publishing, London.

Wheeler, E. A., Baas, P. & Gasson, P. (eds.). (1989). IAWA list of microscopic features for hardwood identification. *IAWA Bulletin* 10: 219–332.

Willemstein, S. C. (1975). Carapa guianensis *Aubl.: a monograph.* Koninklijk Instituut voor de Tropen, Amsterdam.

Wong, T. M. (1976). *Wood structure of the lesser known timbers of Peninsular Malaysia.* Malaysian Forest Records No. 28.

Wong, T. M. (revised by S. C. Lim & R. C. K. Chung) (2002). *A dictionary of Malaysian timbers.* Malaysian Forest Records No. 30. Forest Research Institute, Malaysia.

# INDEX

Numbers in **bold** refer to main taxonomic accounts. The word "mahogany" is only used in the index when qualified, for example "Australian mahogany" or "Caribbean mahogany".

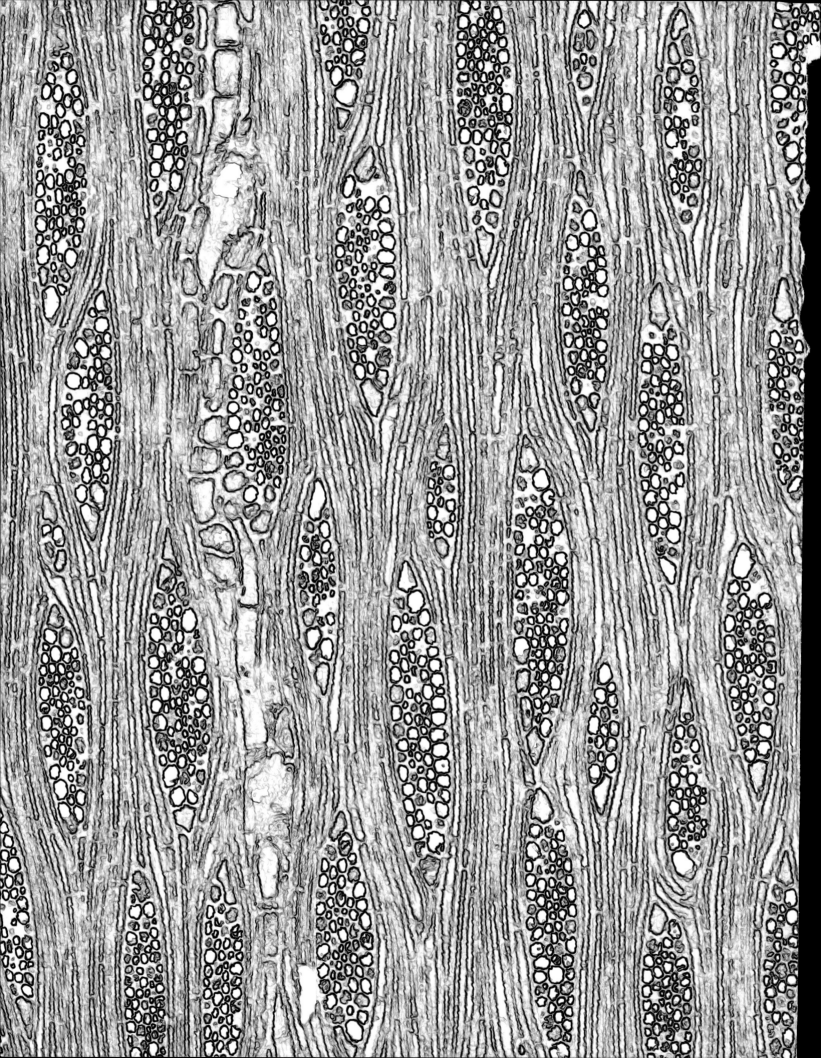